# PRIVATE PROPERTY INFILTRATION AND INFLOW CONTROL

## WEF Special Publication

### 2016

Water Environment Federation
601 Wythe Street
Alexandria, VA 22314-1994 USA
www.wef.org

Private Property Infiltration and Inflow Control

## *About WEF*

The Water Environment Federation (WEF) is a not-for-profit technical and educational organization of 33,000 individual members and 75 affiliated Member Associations representing water quality professionals around the world. Since 1928, WEF and its members have protected public health and the environment. As a global water sector leader, our mission is to connect water professionals; enrich the expertise of water professionals; increase the awareness of the impact and value of water; and provide a platform for water sector innovation. To learn more, visit www.wef.org.

Prepared by the **Private Property Infiltration and Inflow Control**
Task Force of the **Water Environment Federation**

| | |
|---|---|
| Jane McLamarrah, Ph.D., P.E., *Chair* | Jeffrey P. King, P.E., LEED AP, CPESC |
| Aaron Witt, P.E., *Vice-Chair* | Brandon Koltz |
| | George E. Kurz, P.E., DEE |
| Josh Arnold | Ryan C. Laninga |
| Terry A. Bartels, P.E. | Tyler Lewis |
| Mike Beattie | Paul G. Maron, P.E. |
| Gary S. Beck, P.E. | Gary Merriman |
| Kristen E. Buell, P.E., ENV SP | Kate Mowbray |
| | Richard (Rick) E. Nelson, P.E. |
| Paul Burris | Dan Ott, P.E. |
| William C. Carter, P.E. | Christopher Pawlowski, Ph.D., P.E. |
| Laurie Chase, P.E. | |
| David Cooley, P.E. | Robert A. Pennington, P.E., BCEE |
| Darryl Corbin, P.E. | |
| Jaime M. Davidson, P.E. | Charles Poskas, P.E. |
| John Farnkopf, P.E. | James R. Rabine, P.E. |
| Sean W. FitzGerald, P.E. | Christopher Ramsey, P.E. |
| M. Truett Garrett, Jr., P.E., Sc.D. | Matthew Richardson |
| | Jeff Scarano |
| Rhoda Hall, EPI | Nancy Schultz, P.E., D.WRE |
| Eric M. Harold, P.E., BCEE | Russ Stammer, P.E., LEED Green Associate |
| John D. Hendron, P.E. | |
| Stephen P. James, P.E. | David C. Tipping |
| Stephen Jeffus, P.E. | Christina Willson, P.E. |
| Jacqueline Zipkin, P.E. | Michel Wanna, PMP |
| Jack Keys | Tina Wolff, P.E. |

Under the Direction of the **Collection Systems Subcommittee** of the **Technical Practice Committee**

# Contents

List of Figures                                                              xi

List of Tables                                                               xiii

Preface                                                                      xv

**Chapter 1      Introduction**                                              1
*Jaime M. Davidson, P.E.*

1.0   BACKGROUND                                                             1
2.0   OBJECTIVES                                                             3
3.0   ADDITIONAL PRIVATE PROPERTY-RELATED EFFORTS                            4
4.0   REFERENCES                                                             4

**Chapter 2      Private Infiltration and Inflow Definitions and
Problems**                                                                   5
*Jane McLamarrah, Ph.D., P.E.*

1.0   INTRODUCTION                                                           5
2.0   PUBLIC VERSUS PRIVATE RESPONSIBILITIES                                 6
    2.1   Definition of Terms                           6
    2.2   Public Versus Private Responsibilities and Obligations   7
    2.3   Survey Summary                                 8
3.0   EXTENT OF PRIVATE INIFILTRATION AND INFLOW                             10
    3.1   Infiltration and Inflow Sources                10
    3.2   Infiltration and Inflow Magnitude              12
4.0   PRIVATE INFILTRATION AND INFLOW EFFECTS                                14
    4.1   Economic Issues                               14
    4.2   Environmental and Public Health Issues        15
    4.3   Regulatory Issues                             15
5.0   PRIVATE INFILTRATION AND INFLOW
ENFORCEMENT EFFORTS                                                          16
6.0   REFERENCES                                                            17

**Chapter 3      Control of Private Infiltration and Inflow**      19
*William C. Carter, P.E.*

1.0  IDENTIFYING PRIVATE INFILTRATION AND INFLOW
SOURCES      20

  1.1  Measuring Infiltration and Inflow from the Private
Sector      21

    1.1.1  *Flow Monitoring Methods*      21

      1.1.1.1  *Basin Monitoring*      21

      1.1.1.2  *Manual Manhole Service Lateral
Monitoring*      21

      1.1.1.3  *Service Lateral Monitoring*      22

    1.1.2  *Estimating the Private Infiltration and Inflow
Magnitude*      22

    1.1.3  *Establishing the Types of Private Infiltration
and Inflow Sources*      23

  1.2  Private Infiltration and Inflow Source Identification      23

    1.2.1  *Smoke Testing*      23

    1.2.2  *Building Inspections*      24

    1.2.3  *Dyed Water Testing*      25

    1.2.4  *Service Lateral Closed-Circuit Television*      25

      1.2.4.1  *Cleanout Entry Closed-Circuit
Television*      25

      1.2.4.2  *From Main Sewer Closed-Circuit
Television*      26

      1.2.4.3  *Electro-Scan Testing*      26

  1.3  Establishing Infiltration and Inflow Source Flowrates      27

    1.3.1  *Source Runoff Method*      27

    1.3.2  *Typical Private Infiltration and Inflow
Source Rates (Default Flow Method)*      27

    1.3.3  *Field Verification of Source Flows Method*      28

      1.3.3.1  *Dyed Water Testing*      28

      1.3.3.2  *Data Loggers on Sump Pumps*      29

  1.4  Private Sector Infiltration and Inflow Source Data
Management      29

    1.4.1  *Geographical Information System Utilization*      29

    1.4.2  *Asset Management Software*      30

|  |  | 1.4.2.1 | *Building Inspection Modules* | 30 |
|  |  | 1.4.2.2 | *Lateral Modules* | 30 |
| 2.0 | PRIVATE INFILTRATION AND INFLOW CORRECTIVE ACTIONS | | | 30 |
|  | 2.1 | Preventive Design Methods | | 31 |
|  |  | 2.1.1 | *New Sewer Design for Buildings* | 31 |
|  |  | 2.1.2 | *Private Connections to Existing Lines* | 31 |
|  |  | 2.1.3 | *Design Considerations for the Effects of Best Management Practices on Laterals* | 32 |
|  | 2.2 | Corrective Design Methods | | 33 |
|  |  | 2.2.1 | *Cleanout Repairs* | 33 |
|  |  | 2.2.2 | *Downspouts, Driveway Drain, and Area Drain Removals* | 33 |
|  |  | 2.2.3 | *Stairwell Drain Removals* | 34 |
|  |  | 2.2.4 | *Sump Pumps and Foundation Drain Removals* | 34 |
|  |  | 2.2.5 | *Redirecting Private Infiltration and Inflow Source Controls* | 35 |
|  | 2.3 | Private Lateral Repair Methods | | 35 |
|  |  | 2.3.1 | *Complete Liner* | 35 |
|  |  | 2.3.2 | *Partial Liner* | 35 |
|  |  | 2.3.3 | *Grouting* | 36 |
|  |  | 2.3.4 | *Replacement* | 36 |
|  |  |  | 2.3.4.1   *Open Cut* | 36 |
|  |  |  | 2.3.4.2   *Pipe Bursting* | 36 |
|  | 2.4 | Private Infiltration and Inflow Removal Programs | | 37 |
|  | 2.5 | Public Information, Education, and Guidance Initiatives | | 37 |
| 3.0 | PRIVATE INFILTRATION AND INFLOW REMOVAL EFFECTIVENESS | | | 38 |
|  | 3.1 | Preconstruction and Postconstruction Flow Monitoring | | 38 |
|  | 3.2 | Preconstruction and Postconstruction Private Infiltration and Inflow Source Testing | | 38 |
|  | 3.3 | Private Infiltration and Inflow Removal Data Needs | | 38 |
| 4.0 | REFERENCES | | | 39 |
| 5.0 | SUGGESTED READING | | | 39 |

Chapter 4    Private Property Program Implementation
             Considerations                                          41
             *Laurie Chase, P.E.*

1.0   STAKEHOLDER INVOLVEMENT                                         42
      1.1   Customer Stakeholders                                    44
      1.2   Internal Stakeholders                                    44
      1.3   Local Plumbers, Service and Repair Contractors,
            and Builders                                             44
      1.4   Real Estate Industry Interests                           45
      1.5   Regional and Multijurisdictional Programs                45
      1.6   Other Stakeholders                                       45
2.0   PROGRAMMATIC ELEMENTS                                          46
      2.1   Program Scope and Vision                                 46
      2.2   Legal Authorities                                        47
            2.2.1   *Sewer Use Ordinances*                           47
            2.2.2   *Construction Standards and Specifications*      49
      2.3   Program Management and Staffing                          49
            2.3.1   *Management*                                     49
            2.3.2   *Staffing Needs*                                 50
      2.4   Public Education and Communication                       50
      2.5   Budgeting                                                51
      2.6   Information Management                                   51
      2.7   Standard Practices and Acceptable Technologies           52
      2.8   Sustainability (Performance Metrics, Adaptive
            Management)                                              52
3.0   FUNDING CONSIDERATIONS                                         52
      3.1   Expenditure of Public Funds on Private Property          53
      3.2   Diversion of Funds from Other Customer Needs             54
      3.3   Effects on Vulnerable Customers                          54
      3.4   Effects on Local Economy                                 55
      3.5   Effects on Inaction                                      55
      3.6   Utility Capital Improvement Funding Requirements         55
      3.7   Funding Mechanisms                                       56
4.0   POLITICAL AND REGULATORY CONSIDERATIONS                        56
      4.1   Customer Equity and Environmental Justice                56

|  | 4.2 | Private Property Rights | 57 |
|  | 4.3 | Local Political Issues | 58 |
|  | 4.4 | Federal, State, and Local Regulations and Enforcement | 58 |
| 5.0 | | PRIVATE PROPERTY INFILTRATION AND INFLOW PROGRAM EXAMPLES | 58 |
|  | 5.1 | Enforcement-Based Program | 59 |
|  | 5.2 | Point-of-Sale Lateral Inspection/Corrective Action Program | 60 |
|  | 5.3 | Utility-Assumed Ownership and/or Operation and Maintenance of Privately Owned Lateral Program | 61 |
|  | 5.4 | Publicly Owned (Lower) Lateral Focused Program | 64 |
| 6.0 | | REFERENCES | 65 |

**Chapter 5** **Private Property Infiltration and Inflow Program Case Studies** 67
*Aaron Witt, P.E.*

| 1.0 | | WATER ENVIRONMENT FEDERATION PRIVATE PROPERTY VIRTUAL LIBRARY | 68 |
| 2.0 | | JOHNSON COUNTY WASTEWATER, KANSAS | 69 |
|  | 2.1 | Utility Background | 69 |
|  | 2.2 | Private Infiltration and Inflow Removal Program | 70 |
|  |  | 2.2.1 *Program Drivers* | 70 |
|  |  | 2.2.2 *Program Characteristics* | 71 |
|  |  | 2.2.3 *Public Outreach* | 72 |
|  |  | 2.2.4 *Program Resources and Tools for Source Identification and Removal* | 73 |
|  |  | 2.2.5 *Program Costs* | 74 |
|  | 2.3 | Program Effectiveness | 74 |
|  | 2.4 | Conclusions | 77 |
| 3.0 | | KING COUNTY, WASHINGTON | 79 |
|  | 3.1 | Utility Background | 79 |
|  | 3.2 | Private Infiltration and Inflow Removal Program | 80 |
|  |  | 3.2.1 *Program Drivers* | 80 |
|  |  | 3.2.2 *Program Characteristics* | 80 |
|  |  | 3.2.3 *Public Outreach* | 80 |

|  |  | | |
|---|---|---|---|
| | 3.2.4 | *Source Identification and Removal* | 81 |
| | 3.2.5 | *Program Resources and Tools* | 81 |
| | 3.2.6 | *Program Costs* | 82 |
| | 3.3 | Program Effectiveness | 82 |
| | 3.4 | Conclusions | 83 |
| 4.0 | | EAST BAY MUNICIPAL UTILITY DISTRICT | 84 |
| | 4.1 | Utility Background | 84 |
| | 4.2 | Private Infiltration and Inflow Removal Program | 84 |
| | | 4.2.1 *Program Drivers* | 84 |
| | | 4.2.2 *Program Characteristics* | 85 |
| | 4.3 | Program Effectiveness and Conclusions | 88 |
| 5.0 | | SUMMARY | 88 |
| 6.0 | | REFERENCES | 88 |

**Appendix   2015 Water Environment Federation Private Property Infiltration and Inflow Survey   91**
*Tyler Lewis*

| 1.0 | SURVEY BACKGROUND | 91 |
|---|---|---|
| | 1.1 Survey Questions | 92 |
| | 1.2 Survey Responses | 92 |
| 2.0 | UTILITY CHARACTERISTICS | 92 |
| 3.0 | SERVICE CONNECTIONS | 96 |
| 4.0 | INFILTRATION AND INFLOW | 103 |
| 5.0 | ENFORCEMENT AND FINANCING | 110 |
| 6.0 | MISCELLANEOUS | 115 |

**Index**                                                         117

# List of Figures

2.1   Typical lateral installation and terminology ............... 7

2.2   Percentage of service taps, lower laterals, and upper
      laterals located and mapped............................ 9

2.3   Typical private I/I entry sources ...................... 10

2.4   Survey estimated private I/I sources ................... 11

2.5   Percentage of total I/I related to building source
      connections, private laterals, and service taps ........... 12

3.1   Positive private property smoke source .................. 24

3.2   Post-cleanout installation and restoration ............... 34

4.1   City of San Bruno POS flow chart ..................... 62

5.1   Johnson County Wastewater pilot I/I reduction by
      strategy area.......................................... 76

5.2   Johnson County Wastewater construction cost by
      strategy area.......................................... 76

5.3   Johnson County Wastewater unit construction cost vs I/I
      reduction by strategy area ............................. 77

5.4   Johnson County Wastewater I/I removal efficiency by
      strategy area.......................................... 78

# List of Tables

2.1     Estimated I/I contribution levels 2015 survey. . . . . . . . . . . . . . 13

3.1     Private I/I inside and outside building sources . . . . . . . . . . . . 28

5.1     Private I/I investigation participation rates. . . . . . . . . . . . . . . 73

5.2     King County Skyway Project rehabilitation quantities
and costs . . . . . . . . . . . . . . . . . . . . . . . . . . . . . . . . . . . . . . . . . 81

# Preface

This publication is intended to update *Control of Infiltration and Inflow in Private Building Sewer Connections*, which was published by WEF in 1999 under a cooperative agreement with the U.S. Environmental Protection Agency. This revised publication provides an overview of infiltration/inflow (I/I) entering through private property connections, including the results of a survey of wastewater utilities conducted by WEF and the Collection Systems Committee. Guidance is provided on methods to identify, locate, and estimate the magnitude of private property I/I. Guidance is also provided on possible corrective action measures that can be undertaken to control private property I/I. The elements of successful private property programs are described. Specific successful programs are discussed and detailed case studies of leading peer utilities are included.

This publication was produced under the direction of Jane McLamarrah, Ph.D., P.E., *Chair*, and Aaron Witt, P.E., *Vice-Chair*.

Authors' and reviewers' efforts were supported by the following organizations:

AECOM, Cincinnati, Ohio
Arcadis, Fort Wayne, Indiana, and Arlington, Virginia
Blue Heron Engineering Services, Ltd., Dublin, Ohio
Brown and Caldwell, Walnut Creek, California
Carollo Engineers, Inc., Dallas, Texas
CDM Smith, Edison, New Jersey
Civil Design, Inc., St. Louis, Missouri
CH2M, Milwaukee, Wisconsin, and Kansas City, Missouri
East Bay Municipal Utility District, Oakland, California
Fort Wayne City Utilities, Fort Wayne, Indiana
GBA Architects and Engineers, Lenexa, Kansas
GRW, Nashville, Tennessee
Hampton Roads Sanitation District, Newport News, Virginia
HF&H Consultants, LLC, Walnut Creek, California
I+S Group, Faribault, Minnesota
JM Davidson Engineering, D.P.C., North Tonawanda, New York
Johnson County Wastewater, Olathe, Kansas
King County, Washington
J-U-B Engineers, Inc., Coeur d'Alene, Idaho
Losli Engineering, LLC, St. Louis, Missouri

Macon Water Authority, Macon, Georgia

M. T. Garrett and Associates, LLC, Houston, Texas

MWH Americas, Inc., Clemson, South Carolina, and Broomfield, Colorado

New Mexico Environment Department—Surface Water Quality Bureau, Santa Fe, New Mexico

RH2 Engineering, Inc., Bothell, Washington

RJN Group, Dallas, Texas

Strand Associates, Inc., Madison, Wisconsin

U.S. EPA, Washington, District of Columbia

United Water, Grand Rapids, Michigan

Wade Trim, Grand Rapids, Michigan

Water & Environmental Consulting, LLC, Milwaukee, Wisconsin

Wendel Companies, Williamsville, New York

# 1

# Introduction

*Jaime M. Davidson, P.E. and Ryan C. Laninga*

| | | | | |
|---|---|---|---|---|
| 1.0 BACKGROUND | 1 | 3.0 ADDITIONAL PRIVATE PROPERTY-RELATED | | |
| 2.0 OBJECTIVES | 3 | EFFORTS | 4 | |
| | | 4.0 REFERENCES | 4 | |

## 1.0 BACKGROUND

The U.S. Environmental Protection Agency (U.S. EPA) estimates that between 23,000 and 75,000 sanitary sewer overflow (SSO) events occur per year in the United States (excluding basement backups) (U.S. EPA, 2004). Regulatory authorities are increasingly requiring wastewater utilities to eliminate such overflow events, including clarifying requirements for "proper operation and maintenance of the collection system". Aging and deterioration of existing sewer components, as well as poor construction methods in both newer and older collection systems, can lead to cracks and openings that allow the entry of extraneous water (infiltration) to the system. Combined with inflow from directly connected stormwater sources (i.e., downspouts, illicit connections to the sanitary sewer), infiltration and inflow (I/I) can consume valuable sewer capacity in both gravity pipes and pumping stations. Ultimately, excess I/I can add to conveyance and treatment costs for all utility customers, contribute to SSOs and combined sewer overflows (CSOs) from wastewater collection systems, and potentially result in enforcement orders from either state or federal regulatory agencies.

Based on a limited sample of six communities, U.S. EPA estimated that 27% of all SSOs are caused by I/I and another 7% are caused by insufficient system capacity (U.S. EPA, 1996). In the U.S. EPA CSO/SSO report to the U.S. Congress, SSO cause information was available for 77% of the SSO events included in the SSO data management system. Of those events with a known cause, 26% of the events were the result of wet weather and I/I (U.S. EPA, 2004). Thus, one approach to eliminating SSOs and CSOs is to reduce the entry of I/I into the sewer system.

Utilities that attempted to mitigate I/I in their collection system began to recognize that I/I control efforts are a significant challenge. As cracks and leaks in one part of the system were corrected, groundwater was frequently able to migrate through bedding to the next weakest part of the collection system, which may comprise an adjacent pipe segment or a private sewer connection. Although SSO reduction efforts have historically been focused on public sewer systems, the privately owned portions of the sewer system have the potential to significantly contribute to the number of overflow events. In some cities, it is estimated that as much as 60% of the flow that overfills sanitary sewers comes from service connections (U.S. EPA, 1996).

If removing public I/I sources alone is not enough, wastewater utilities are faced with the dilemma of either constructing large storage structures, when permitted by regulatory agencies, or going onto private property to remove private I/I sources. More often, utilities and regulatory agencies recognize a need to minimize I/I entry into the collection system in a holistic way that addresses both public collection system components and private sources of I/I from customers rather than store extraneous flow.

There are many roadblocks, however, that prevent utilities from implementing a private property program. For instance, many communities do not know where to begin. The quantity of private sewer connections can be overwhelming to monitor and isolate from the main collection system. Moreover, it is often difficult to work on private property because of legal constraints, easement issues, and ownership uncertainties. Funding the program is another common challenge because of limited budgets, staffing, and resources to undertake repairs, as well as the question of spending public money on private property. Additionally, it can be difficult to gain stakeholder support to initiate the programs.

In 1999, the Water Environmental Federation (WEF), in conjunction with U.S. EPA, conducted a national survey to gather information on wastewater collection systems, service connections, and the way municipalities are addressing I/I problems. The results of the survey were published in a monograph entitled *Control of Infiltration and Inflow in Private Building Sewer Connections* (Sanitary Sewer Overflow Cooperative Agreement Workgroup of the Water Environment Federation, 1999). In the 15 years since the monograph was published, management of private property-related I/I has continued to be at the forefront of discussions for utilities nationwide. As a result, in 2015, WEF and the authors of this publication conducted an Internet survey entitled, "2015 WEF Private Property I/I Survey", to update the information originally presented in the 1999 monograph. The results of the 2015 WEF Private Property I/I Survey are presented in Chapter 2, and a summary of the data collected is included in the appendix of this chapter.

As confirmation of the magnitude of I/I contribution from private property, 31% of the respondents to the 2015 WEF Private Property I/I Survey estimated that private I/I sources contribute 50 to 70% of their total I/I, while 36% estimated a contribution of 20 to 50%; only 8% estimated that less than 20% is from private I/I contribution.

## 2.0  OBJECTIVES

This publication is intended to serve as a resource for utilities that either have a private property program in place or are looking to establish one to address private property-related I/I within their collection system. Specifically, the authors aim to

- Define common terminology related to private property sewer connections;
- Present the information collected through the 2015 WEF Private Property I/I Survey, including how private property I/I affects collection system operators of varying size nationwide and how these operators have been working to address the issue;
- Describe the methodologies available to investigate and remediate private sewer defects;
- Give utilities an understanding of the legal, funding, political, and regulatory considerations that need to be addressed as part of a private property program;
- Outline the types of private property programs that utilities nationwide have implemented, including enforcement-based programs, point-of-sale inspection programs; and assumption of public responsibility for maintenance of varying portions of the private sewer; and
- Share case studies on successful private property programs.

As utilities nationwide face similar challenges related to private property, it is evident that sharing resources on both successful and unsuccessful approaches to the issue can be beneficial to everyone.

## 3.0 ADDITIONAL PRIVATE PROPERTY-RELATED EFFORTS

Following the 1999 monograph publication, WEF conducted a Webcast on private property issues in August 2005 that was the highest viewed Webcast WEF had sponsored to date. Participants requested example ordinances, letters, and other documents from speakers, which led the WEF Collection Systems Committee to begin the Private Property Virtual Library (PPVL) project to collect and share this type of information. The original project objective was to develop a virtual library of private property program-related resource materials to be available through the Internet. Initial project efforts focused on "successful" private property programs targeting sanitary lateral repair or replacement, I/I source detection and elimination, and lateral condition assessment. The PPVL, which can be accessed at www.wef.org, is discussed in more detail in Chapter 5.

## 4.0 REFERENCES

Sanitary Sewer Overflow Cooperative Agreement Workgroup of the Water Environment Federation (1999) *Control of Infiltration and Inflow in Private Building Sewer Connections;* Water Environment Federation: Alexandria, Virginia.

U.S. Environmental Protection Agency (2004) *Report to Congress on the Impacts and Control of CSOs and SSOs.* http://www.epa.gov/npdes/2004-npdes-cso-report-congress (accessed Jan 2016).

U.S. Environmental Protection Agency (1996) *Sanitary Sewer Overflows: What Are They and How Can We Reduce Them?* http://www3.epa.gov/npdes/pubs/ssodesc.pdf (accessed Jan 2016).

# 2

# Private Infiltration and Inflow Definitions and Problems

*Jane McLamarrah, Ph.D., P.E. and Tyler Lewis*

| | | |
|---|---|---|
| 1.0 INTRODUCTION | 5 | 4.0 PRIVATE INFILTRATION AND INFLOW EFFECTS 14 |
| 2.0 PUBLIC VERSUS PRIVATE RESPONSIBILITIES | 6 |    4.1 Economic Issues 14 |
|    2.1 Definition of Terms | 6 |    4.2 Environmental and Public Health Issues 15 |
|    2.2 Public Versus Private Responsibilities and | |    4.3 Regulatory Issues 15 |
|       Obligations | 7 | 5.0 PRIVATE INFILTRATION AND INFLOW |
|    2.3 Survey Summary | 8 |    ENFORCEMENT EFFORTS 16 |
| 3.0 EXTENT OF PRIVATE INFILTRATION AND INFLOW | 10 | 6.0 REFERENCES 17 |
|    3.1 Infiltration and Inflow Sources | 10 | |
|    3.2 Infiltration and Inflow Magnitude | 12 | |

## 1.0 INTRODUCTION

As discussed in Chapter 1, in 1999, the Water Environmental Federation (WEF), in conjunction with the U.S. Environmental Protection Agency (U.S. EPA), conducted a national survey to gather information on wastewater collection systems, service connections, and the way municipalities are addressing infiltration and inflow (I/I) problems. The results of the survey were published in a monograph entitled, *Control of Infiltration and Inflow in Private Building Sewer Connections* (Sanitary Sewer Overflow Cooperative Agreement Workgroup of the Water Environment Federation, 1999). This publication updates the 1999 monograph based on the 2015 WEF Private Property I/I Survey conducted by WEF and the authors.

Chapter 2 defines the terminology and discusses public vs private infrastructure responsibilities, discusses estimates of the magnitude of private I/I, and describes potential effects of private I/I. The results of the 2015 WEF Private Property I/I Survey are referenced throughout to provide respondents' experiences.

# 2.0   PUBLIC VERSUS PRIVATE RESPONSIBILITIES

One of the key obstacles utilities face in controlling I/I entering public sewer mains from private property sources is the lack of consistent definitions in terminology and in where the public vs private ownership and operation and maintenance (O&M) responsibilities reside (i.e., at the tap on the sewer main, at the property or easement line, or at the connection to the building plumbing). Section 2.1 in this chapter defines terminology used in this publication. Section 2.2 describes the variation in public vs private responsibilities. Section 2.3 summarizes the 2015 WEF Private Property I/I Survey. The full 2015 WEF Private Property I/I Survey and survey results are included in the Appendix of this publication.

## 2.1   Definition of Terms

The industry-standard terms for publicly owned gravity sewer pipes are "sewer mains" and "mainline sewers". Here, however, the consistency ends. Different terms are used in different geographic regions for sewer pipes connecting the customer's building to the sewer main, including "building sewer connection", "building service line", "side sewer", and "lateral". Increasing consistency seems to be arising for the use of the term, "lateral", with "upper lateral" meaning the pipe from the building to the street right-of-way or easement line and "lower lateral" meaning the remaining pipe to the connection point at the sewer main. Figure 2.1 provides a typical installation of the connections from house plumbing to the public sewer main and graphically illustrates the terminology used in this publication. Where future references in this publication refer to both the upper and lower laterals, the term, "entire lateral", will be used.

Most utilities (i.e., 43% of the sample population of 47 respondents from the 2015 WEF Private Property I/I Survey) require installation of a building cleanout near the building to connect the building plumbing and the upper lateral, providing a necessary access point for clearing potential blockages from the pipe. Thirty-eight percent of the surveyed utilities indicated that

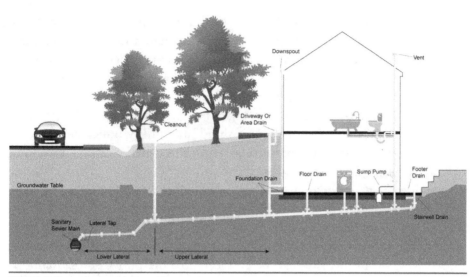

**FIGURE 2.1**   Typical lateral installation and terminology.

building cleanout typically exists. Many utilities (i.e., 35% of those surveyed) also require installation of a cleanout at the right-of-way line. A cleanout at the right-of-way line delineates the division between the upper and lower lateral pipes and provides additional cleanout access and the ability to determine if a blockage is in the lateral pipe from the building vs the public sewer, typically in the street right-of-way; 22% of the survey utilities indicated that the right-of-way cleanout typically exists. When the public sewer is located in a backyard or side-yard easement, 19% of the surveyed utilities require a cleanout at the easement line, with 11% indicating that the easement cleanout typically exists. As seen from the survey, more utilities are requiring right-of-way, and, for somewhat fewer utilities, easement cleanouts. Utilities such as Dallas Water Utilities have had programs to retrofit such cleanouts for existing customers without original cleanout installations.

## 2.2   Public Versus Private Responsibilities and Obligations

Utility O&M responsibilities for laterals vary. Approximately 32% of the surveyed utilities consider only the sewer main in the street, excluding the tap, to be a public asset. Approximately 18% of the surveyed utilities assume public responsibility for the tap at the lateral connection to the sewer main, excluding the lateral pipe and cleanouts. A majority of utilities (i.e., 47% of the surveyed utilities) assume responsibility for the tap and the lower lateral, typically including the street/easement cleanout. Only 3% of the surveyed utilities assume responsibility for the tap, the lower lateral, and the upper lateral all the way to the building cleanout.

Utility ownership responsibilities generally follow the same limits as the aforementioned O&M responsibilities. At times, a utility may assume additional operational responsibilities to provide good customer service. Approximately 17% of the surveyed utilities have assumed operational responsibilities, while leaving ownership and maintenance responsibility with the property owner. Another 17% indicated having legal authority to assume operational responsibilities should the utility policy change. An example of an additional operational service is when a utility provides blockage-clearing services for all or a portion of the privately owned lateral, but requires the property owner to perform any necessary maintenance such as structural repairs, rehabilitation, or replacement that may be needed for the privately owned portions of the lateral.

Most utilities also have sewer use regulations defining the types of building fixtures and drains that can be connected to the public sewer main or manhole (some utilities only allow connections at manholes) to control the sources of discharge into the sewer main. More than 94% of the surveyed utilities maintain written procedures and technical specifications for the proper connection of a building to a public sewer line, with only 6% of the respondents noting no such procedures or specifications. Legal authorities governing private laterals, building plumbing connections, and source control requirements are detailed in Chapter 4 ("Private Property Program Implementation Considerations").

## 2.3    Survey Summary

As noted in the introduction to this chapter, a 2015 Internet survey was completed to gather current information on private property I/I conditions. An e-mail invitation to the survey was distributed to all WEF utility members. Forty-seven respondents voluntarily completed the survey. This 2015 WEF Private Property I/I Survey response from 47 utilities was significantly less than the original 1999 survey, which distributed 3,108 questionnaires and had 316 completed, or partially completed, questionnaires returned. Tables and figures with key information from the 2015 WEF Private Property I/I Survey are included in various chapters of this publication as topic appropriate, with the complete survey results included in the Appendix.

The majority of the respondents (i.e., 89%) were municipal or other governmental agencies, with another 9% representing special-purpose districts and the remaining 2% comprising private and investor-owned utilities. One consultant and one state regulatory agency also completed sections of the survey. Utility size ranged from approximately 8.5 mi of separated sewer to more than 2200 mi of separated sewer. Both separate and combined sewer systems were included, with all responding utilities indicating separate sewers and six respondents noting combined sewers, ranging from 5 to 219 mi of combined sewer. Responding utilities also noted systems with grinder pumps (23 respondents), vacuum sewers (one respondent), and septic tank effluent pump (STEP) system components (eight components) (one respondent noted not owning the grinder pumps, but owning the pressure sewers serving the grinder pump installations; one regulatory respondent also noted that the state of Kansas has more than 500 gravity systems, about 10 pressure systems, and three STEP systems).

As shown in Figure 2.2, the surveyed utilities generally know the location of service taps on the public sewers, with roughly 68% having located and mapped more than one-half of the active service taps. The percentage of survey utilities having located and mapped more than one-half of the lower laterals in their system goes down to approximately 57% and, for more than one-half of the upper laterals, to 38%.

Inspection programs for service taps, lower laterals, and upper laterals show similar higher rates for surveyed utilities for service tap inspections and declining rates for lower laterals and upper laterals. As expected, the highest inspection programs are for new installations, with fewer programs

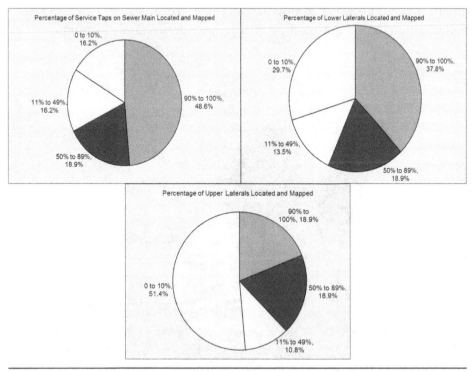

**FIGURE 2.2**  Percentage of service taps, lower laterals, and upper laterals located and mapped.

for repair activities and even fewer programs during I/I or other investigations. Relatively few surveyed utilities (i.e., 3% for service taps, 0% for lower laterals, and 5% for upper laterals) have inspection programs upon transfer of property ownership.

## 3.0    EXTENT OF PRIVATE INFILTRATION AND INFLOW

While it is difficult to quantify I/I entering wastewater collection systems, it is especially difficult to segregate the quantity of I/I entering from private sources. Section 3.1 identifies the sources of private I/I and Section 3.2 discusses estimates of the amount of private I/I.

### 3.1    Infiltration and Inflow Sources

The typical sources of I/I entry into the private lateral and appurtenances are depicted in Figure 2.3. Infiltration enters private laterals through pipe cracks and joints, especially where the "holes" are made larger as roots enter the pipe and grow in size; improperly installed cleanouts; poorly installed connections to the sewer main; pipe failures from traffic overburden or loss of supporting soil; and leaking joints. Inflow enters private laterals through intentional connections of downspouts (e.g., roof drains), sump pumps, building foundation drains, area or driveway drains, and cleanouts.

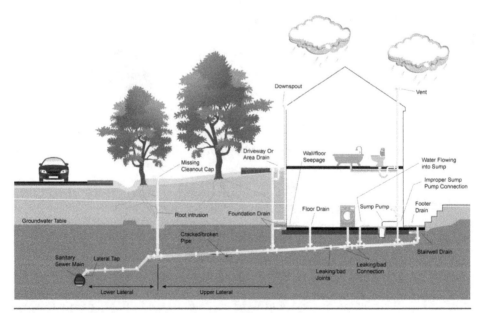

**FIGURE 2.3**    Typical private I/I entry sources.

Private lateral I/I sources may exist even when the lateral is new if the facilities were installed improperly and inspectors failed to recognize substandard performance or illegally connected cross connections. Frequently, cross connections installed during initial building construction were legal at the time of connection, but have since become illegal because of code changes to reduce the amount of stormwater entering the wastewater system. In some systems, subdivisions were commonly graded to direct surface water and wall drainage into depressed backfill at the foundation, with foundation drains directly connected to pervious sewer trenches (CH2M Hill Engineering Ltd., 1994). Survey respondents were asked to rate the highest private I/I sources from 1 (the highest entry source) to 4 (the lowest entry source). As shown from the results in Figure 2.4, building sources and upper lateral sources tied for being the highest contributing source, followed closely by lower lateral sources.

Private I/I sources also frequently increase as the lateral ages, especially when no maintenance or inspection is performed. Lateral and cleanout failures start with cracking, deflection, crown sag, offset joints, deteriorated mortar, and exposed reinforcing. Factors contributing to these failures include size of the defect, soil type, trench-bedding characteristics, sewer sedimentation, root intrusion, groundwater level and fluctuation, internal and external corrosion, method of construction, and external loading.

Private I/I entry can actually increase because of improvements in the adjacent collection system. Some utilities have completed I/I control projects to repair, rehabilitate, and replace public sewers and have found that localized groundwater levels began to rise because groundwater could no

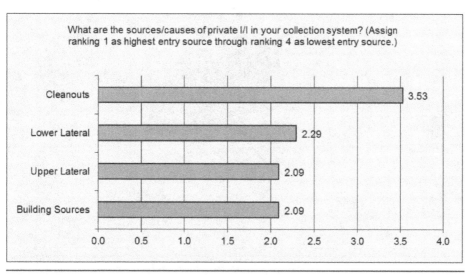

FIGURE 2.4  Survey estimated private I/I sources.

longer enter the sewers. The increase in groundwater elevation allowed the water to enter the typically shallower private lateral pipes and re-enter the collection system through the private laterals.

While the primary focus of this publication is I/I entry through private laterals, significant sources of private I/I entering collection systems can be privately owned collection systems. Private collection systems may be installed to serve such customers as apartment complexes, commercial shopping centers, or industrial facilities.

## 3.2    Infiltration and Inflow Magnitude

There are relatively few utilities that have expended the effort and resources needed to quantify the amount of I/I entering the collection system through private sources. It can be difficult to quantify private I/I sources because of the number of private connections to the collection system. Installing flow meters is cost prohibitive and is technologically challenging to measure relatively low flows from individual lateral connections. Thus, much of the data on the extent of private I/I entry are estimated from typical individual source entry rates (in gallons per minute) multiplied by the projected number of entry sources (from pilot program inspections or sewer system evaluation survey field work). The 2015 WEF I/I Survey results described herein demonstrate the relatively limited quantifiable knowledge of private I/I entry.

The respondents' estimates of the percentage of total I/I related to building source connections, private laterals, and service taps are summarized in Figure 2.5. As shown, 31% of the respondents note private I/I sources as

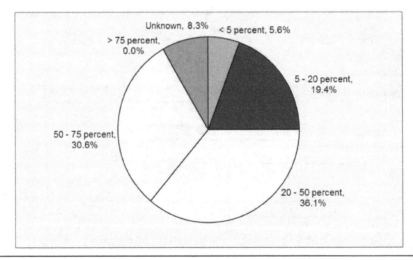

**FIGURE 2.5**   Percentage of total I/I related to building source connections, private laterals, and service taps.

contributing 50 to 75% of the I/I, and 36% of the respondents note contributing 20 to 50%. Only 25% of the respondents estimate less than a 20% private I/I contribution rate. Approximately 8% of the respondents noted that the private I/I contribution rate was unknown. However, 61% of the respondents to the private I/I contribution rate indicated the provided estimate was "a guess". Approximately 27% estimated private I/I contribution rates based on source defect unit contribution rates and approximately 12% estimated private I/I contribution rates based on basin flow monitoring.

The survey also asked respondents to characterize I/I contributions for a dry year, for an average year, and for a wet year. A comparison of how the dry year contributions change during average and wet years is provided in Table 2.1. During a dry year, none of the respondents estimated I/I contributions greater than 40% and only two utilities estimated average year contributions.

One utility that has expended resources to quantify private I/I contributions is the City of Columbus, Ohio. As part of this study, the city performed detailed drainage and connectivity investigations of 116 private houses located in the Barthman-Parsons area, which is served by a separated sewer system. Private residential sources of I/I were identified and then estimates of I/I contributions from residential properties in different neighborhood clusters were developed. These estimated I/I contributions were extrapolated to the entire Barthman-Parsons area. These residential sources were estimated to make up approximately 35% of total I/I for short, intense storms with dry antecedent conditions and approximately 7% of the total I/I under low-intensity, long-duration storms with wet antecedent conditions (Pawlowsik et al., 2014).

**TABLE 2.1**   Estimated I/I contribution levels per 2015 survey.

| Dry year I/I contribution levels | Number of respondents | Number of utility respondents | | | | | | | | | |
| --- | --- | --- | --- | --- | --- | --- | --- | --- | --- | --- | --- |
| | | Average year I/I | | | | | Wet year I/I | | | | |
| | | Unsure | <20% | 20 to 40% | 40 to 60% | 60 to 80% | Unsure | <20% | 20 to 40% | 40 to 60% | 60 to 80% |
| Unsure | 3 | 3 | 0 | 0 | 0 | 0 | 3 | 0 | 0 | 0 | 0 |
| <20% | 29 | 0 | 9 | 20 | 0 | 0 | 2 | 2 | 10 | 13 | 2 |
| 20 to 40% | 3 | 0 | 0 | 1 | 2 | 0 | 0 | 0 | 0 | 1 | 2 |
| Totals | 35 | 3 | 9 | 21 | 2 | 0 | 5 | 2 | 10 | 14 | 4 |

## 4.0    PRIVATE INFILTRATION AND INFLOW EFFECTS

Private I/I effects tend to be economic, public health, environmental, or regulatory-related. Each of these types of effects is described in this section.

### 4.1    Economic Issues

Many utilities are only concerned with private I/I when the amount of I/I entering the collection system becomes costly to convey and treat. This was the case for public I/I source correction projects during the Construction Grants Program era, when U.S. EPA only funded the cost-effective removal of public I/I. Infiltration and inflow was considered to be cost-effective to remove if the cost of the corrective action was less than the cost to convey and treat the I/I (U.S. EPA, 1975). While grants under this program have ended, State Revolving Fund loan applicants are still typically required to evaluate the effects of I/I in the overall system. This evaluation typically starts with a screening to compare sewered population to water resource recovery facility (WRRF) flow to determine gallons per capita per day (gpcd). The states' standards for excessive infiltration vary between 100 and 150 gpcd (U.S. EPA, 2014a). Existing U.S. EPA guidance, which uses 120 gpcd, was published in 1985 when 3.5-gal-per-flush toilets were standard (the Energy Policy Act of 1992 required that newly installed toilets use a maximum of 1.6 gal per flush) (U.S. Congress, 1992). U.S. EPA has also cited criteria on excessive I/I that indicate that infiltration rates in excess of 1000 gal/d per inch diameter per mile of pipe are excessive (U.S. EPA, 2014a). As with excessive infiltration, standards for excessive inflow vary. U.S. EPA has cited average wastewater flows (excluding significant industrial or commercial flows) at WRRFs that exceed 275 gpcd to be excessive (U.S. EPA, 2014b).

Determining whether I/I is cost-effective to remove is highly dependent on a particular utility's cost of conveyance and treatment, especially for communities that do not use full life-cycle costs in establishing utility rates. In a community with significant numbers of capacity-limited sewers or with WRRFs having site or effluent discharge constraints for expansion, the cost of conveying and treating I/I is higher and thus more effort may be needed to control I/I at the source. Utilities with fewer capacity constraints may be willing to convey and treat larger quantities of I/I.

In addition to the public costs of conveying and treating I/I, there are also other economic effects for the utility as well as private economic effects. If I/I causes downstream effects such as building backups, sanitary sewer overflows (SSOs), or combined sewer overflows (CSOs), utilities incur costs of responding to and controlling such events. Customers may incur property damages because of building backups or SSOs and, depending on many factors, may file damage claims against the utility. Building backup damage claims may include private-property damages from localized flooding claims or issues associated with rising groundwater elevations that cause water intrusion to buildings.

## 4.2    Environmental and Public Health Issues

In addition to the economic issues associated with building backups, SSOs, and CSOs, such events may also cause environmental and public health problems (U.S. EPA, 2004). Raw wastewater discharges may contain high levels of pathogenic microorganisms, suspended solids, toxic pollutants, floatables, nutrients, oxygen-demanding components, pharmaceutical products, and oil and grease. Such discharges are a risk to public health through contact with raw or diluted wastewater, especially for immune-compromised individuals or individuals with broken skin that may come in contact with the contaminants. Such discharges also increase pollutant loads in receiving waters and threaten aquatic habitats.

## 4.3    Regulatory Issues

Besides the potential adverse public health and environmental effects of SSOs, such events are considered unpermitted discharges under the Clean Water Act (CWA). Combined sewer overflow discharges, while permitted, have similar adverse public health and environmental effects. To the extent that I/I contributes to SSOs and CSOs, the entry of I/I into collection systems can trigger regulatory enforcement actions under the CWA. U.S. EPA enforcement orders have long focused on controlling I/I through closing constructed SSOs; minimizing or eliminating CSOs; requiring peak flow management plans; and addressing capacity-limited sewers, pumping stations, and plants. More recent enforcement orders are adding O&M requirements as suggested by capacity, management, operation, and maintenance (CMOM) guidance documents and, most recently, by including references to controlling private I/I. For example, the Miami-Dade County Consent Decree has a number of CMOM program elements required and defines building backups as SSOs (U.S. District Court for the Southern District of Florida, 2013).

Even in instances where a system is not suffering building backups, SSOs, or CSOs, systems experiencing surcharging or where significant new growth is expected and existing system capacity is inadequate or marginal may be good candidates for further I/I investigation and potential private I/I control.

## 5.0 PRIVATE INFILTRATION AND INFLOW ENFORCEMENT EFFORTS

To implement private I/I source controls, utilities must have adequate legal authority for such programs. This legal authority requires the ability to enter private property to inspect conditions, to enforce provisions of local sewer use ordinances or regulations, and to adopt and spend public monies on such programs. The legalities and legal interpretations vary between states and even between utilities within the same state, but, in general, there is a reluctance throughout the country to expend public monies to improve private property. Such expenditures may be considered illegal "bribery cases" under more egregious cases, or merely undesirable "favoritism cases" when such expenditures are fully disclosed to the public. Some states, such as Texas, prohibit expenditures of public funds for private property in their state constitution.

However, failure to address private I/I sources because of these legal obstacles or because of the difficulty of implementing such controls can have undesirable outcomes for other customers. Just as with collection system facilities, private laterals age and deteriorate. As additional cracks and leaks develop, lateral pipes, joints, and cleanouts are subject to increasing amounts of I/I entry that then must be conveyed and treated by the utility. These conveyance and treatment costs are shared by all customers within the system. Failure to require maintenance or replacement of the aging laterals and appurtenances will only allow even greater amounts of I/I into the collection system in future years.

Similarly, failure to require the disconnection of existing, generally illegal cross connections will allow greater amounts of inflow to the collection system. While the number of cross connections typically remains relatively constant, one of the effects of climate change is that rainfall patterns seem to have more extreme events. Rainfall peaks seem to be higher, and, with the higher peak rain events, more inflow enters the collection system through the cross connections.

The survey respondents overwhelmingly (i.e., 94%) indicated having the legal authority to require disconnection of sump pumps, downspouts, roof drains, and area drain connections to the public sewer system. Approximately 57% of the respondents noted routinely implementing this legal authority when needed. Another 37% indicated this legal authority was only occasionally exercised. Only 6% of the respondents noted not having the legal authority to disconnect such sources.

## 6.0  REFERENCES

CH2M Hill Engineering Ltd., Edmonton (1994) *Lot Drainage Character-istics Study Natural Storm Events;* Alberta Municipal Affairs Innovative Housing Grants Program: Alberta, British Columbia, Canada.

Pawlowski, C. W.; Rhea, L.; Shuster, W. D.; Barden, G. (2014) Some Factors Affecting Inflow and Infiltration from Residential Sources in a Core Urban Case Study in a Columbus, Ohio, Neighborhood. *J. Hydraulic Eng.,* **140** (1).

Sanitary Sewer Overflow Cooperative Agreement Workgroup of the Water Environment Federation (1999) *Control of Infiltration and Inflow in Private Building Sewer Connections;* Water Environment Federation: Alexandria, Virginia.

U.S. Congress (1992) Energy Policy Act of 1992. https://www.ferc.gov/legal/maj-ord-reg/epa.pdf (accessed Feb 2016).

U.S. Environmental Protection Agency (1975) *Handbook for Sewer System Evaluation and Rehabilitation.* http://www.electroscan.com/wp-content/uploads/2013/02/1975_09-01-EPA-Handbook_Sewer-System-Evaluation-and-Rehabilitation_EPA-430-9-75-021.pdf (accessed Feb 2016).

U.S. Environmental Protection Agency (2004) *Report to Congress Impacts and Control of Combined Sewer Overflows and Sanitary Sewer Overflows.* http://www.epa.gov/sites/production/files/2015-10/documents/csossortc2004_full.pdf (accessed Feb 2016).

U.S. Environmental Protection Agency (2014a) *Guide for Estimating Infiltration and Inflow.* U.S. EPA Water Infrastructure Outreach. http://www3.epa.gov/region1/sso/pdfs/Guide4EstimatingInfiltrationInflow.pdf (accessed Feb 2016).

U.S. Environmental Protection Agency (2014b) *Quick Guide for Estimating Infiltration and Inflow for Region 1 NPDES Annual Reporting.* U.S. Water Infrastructure Outreach. http://www3.epa.gov/region1/sso/pdfs/QuickGuide4EstimatingInfiltrationInflow.pdf (accessed Feb 2016).

U.S. District Court for the Southern District of Florida (2013) *Case No. 1:12-cv-24400-FAM,* Miami-Dade County, Florida (Defendant); the State of Florida, the Florida Department of Environmental Protection and the U.S. Environmental Protection Agency (Plaintiffs), Entered June 6, 2013. https://www.miamidade.gov/water/library/reports/consent-decree/consent-decree-signed.pdf (accessed Feb 2016).

# 3

# Control of Private Infiltration and Inflow

*William C. Carter, P.E.; Gary S. Beck, P.E.; and Josh Arnold*

1.0 IDENTIFYING PRIVATE INFILTRATION AND
INFLOW SOURCES 20
  1.1 Measuring Infiltration and Inflow from
      the Private Sector 21
    1.1.1 Flow Monitoring Methods 21
      1.1.1.1 Basin Monitoring 21
      1.1.1.2 Manual Manhole Service
              Lateral Monitoring 21
      1.1.1.3 Service Lateral
              Monitoring 22
    1.1.2 Estimating the Private Infiltration
          and Inflow Magnitude 22
    1.1.3 Establishing the Types of Private
          Infiltration and Inflow Sources 23
  1.2 Private Infiltration and Inflow Source
      Identification 23
    1.2.1 Smoke Testing 23
    1.2.2 Building Inspections 24
    1.2.3 Dyed Water Testing 25
    1.2.4 Service Lateral Closed-Circuit
          Television 25
      1.2.4.1 Cleanout Entry Closed-
              Circuit Television 25

      1.2.4.2 From Main Sewer
              Closed-Circuit
              Television 26
      1.2.4.3 Electro-Scan Testing 26
  1.3 Establishing Infiltration and Inflow
      Source Flowrates 27
    1.3.1 Surface Runoff Method 27
    1.3.2 Typical Private Infiltration and
          Inflow Source Rates (Default
          Flow Method) 27
    1.3.3 Field Verification of Source
          Flows Method 28
      1.3.3.1 Dyed Water Testing 28
      1.3.3.2 Data Loggers on Sump
              Pumps 29
  1.4 Private Sector Infiltration and Inflow
      Source Data Management 29
    1.4.1 Geographical Information System
          Utilization 29
    1.4.2 Asset Management Software 30
      1.4.2.1 Building Inspection
              Modules 30
      1.4.2.2 Lateral Modules 30

2.0  PRIVATE INFILTRATION AND INFLOW
      CORRECTIVE ACTIONS                          30
      2.1  Preventive Design Methods              31
           2.1.1  New Sewer Design for Buildings  31
           2.1.2  Private Connections to Existing
                  Lines                           31
           2.1.3  Design Considerations for the
                  Effects of Best Management
                  Practices on Laterals           32
      2.2  Corrective Design Methods              33
           2.2.1  Cleanout Repairs               33
           2.2.2  Downspout, Driveway Drain,
                  and Area Drain Removals        33
           2.2.3  Stairwell Drain Removals       34
           2.2.4  Sump Pumps and Foundation
                  Drain Removals                 34
           2.2.5  Redirecting Private Infiltration
                  and Inflow Source Flows        35
      2.3  Private Lateral Repair Methods        35
           2.3.1  Complete Liner                 35

           2.3.2  Partial Liner                  35
           2.3.3  Grouting                       36
           2.3.4  Replacement                    36
                  2.3.4.1  Open Cut              36
                  2.3.4.2  Pipe Bursting         36
      2.4  Private Infiltration and Inflow Removal
           Programs                              37
      2.5  Public Information, Education, and
           Guidance Initiatives                  37
3.0  PRIVATE INFILTRATION AND INFLOW REMOVAL
      EFFECTIVENESS                              38
      3.1  Preconstruction and Postconstruction
           Flow Monitoring                       38
      3.2  Preconstruction and Postconstruction
           Private Infiltration and Inflow Source
           Testing                               38
      3.3  Private Infiltration and Inflow Removal
           Data Needs                            38
4.0  REFERENCES                                  39
5.0  SUGGESTED READING                           39

Control of private property infiltration and inflow (I/I) is necessary for both new and old systems. This chapter presents methods that have been used to identify private I/I sources and methods for the repair and protection of systems from private I/I.

## 1.0  IDENTIFYING PRIVATE INFILTRATION AND INFLOW SOURCES

A step-by-step process is typically conducted to identify private sector I/I sources. The step-by-step process presented in this section has been conducted for many years. The process has been enhanced by new approaches and equipment used to identify private sector I/I sources.

## 1.1    Measuring Infiltration and Inflow from the Private Sector

Flow monitoring is typically the first step conducted to establish I/I magnitudes in a system. If the I/I magnitude is excessive, then removal of I/I sources may be prudent. Flow monitoring data can indicate that private sector I/I may be excessive and where to begin search efforts to identify I/I sources. The following sections provide insight to measuring private sector I/I in sanitary sewer systems.

### 1.1.1    Flow Monitoring Methods

Flow monitoring equipment has improved over the years and is becoming more prevalent as a tool to indicate the health of sanitary sewer systems as they relate to I/I. Although identifying I/I from flow data is relatively straightforward, it is difficult to separate private sector I/I (i.e., building sources and service laterals) from I/I that comes from the public sector (i.e., mains and manholes). The following are current and new practices for flow monitoring to establish where private I/I may be excessive and specific methods of measuring I/I from the private sector.

#### 1.1.1.1    Basin Monitoring

Flow monitoring at the basin level is typically used to establish the levels of I/I within that basin, which is then compared with a guideline level to define whether I/I is "excessive" in that basin. Sanitary sewer basins generally range in size from 1.5 km (5000 ft) of sewer to 30.5 km (100 000 ft) of sewer. The basin should not be so large as to lose consistent system characteristics and not too small to have difficulty measuring base dry weather flows.

Excessiveness of I/I in a basin, defined in Section 4.1 of Chapter 2 (entitled, "Economic Issues"), can be determined in many ways. One method is to establish system capacity and to determine if the peak wet weather flows from the basin are causing problems in the system such as backups or overflows. Another method is to compare the I/I levels to typical I/I flows from similar basin types. If I/I flows are determined to be excessive in a basin, then the sewer authority needs to establish which investigative methods are required (i.e., smoke testing, building inspection, or others).

#### 1.1.1.2    Manual Manhole Service Lateral Monitoring

Manual manhole service lateral monitoring is a specific method for measuring I/I from the private sector. It is conducted by measuring flows entering manholes from service laterals that are directly connected to the manhole. The process typically includes a cup or bucket with liquid gradations (i.e., ounces) and a stopwatch to determine a rate of flow. Testing is typically conducted during or immediately after a rainfall event. This method of testing is typically used to determine the flowrates for several selected laterals in a basin or system. From this data, a typical lateral I/I flowrate is established and used to project the I/I associated with all laterals in the system. Follow-up inspections of the service laterals are conducted during dry weather to determine if the lateral is typically dry and to make sure that the lateral flow measured during the storm events was not a typical dry weather flow.

### 1.1.1.3 Service Lateral Monitoring

Recently, attempts have been made to measure lateral flow and to identify wet weather flow volumes using down-facing ultrasonic depth sensors mounted in the service lateral cleanout. This is a specific method for measuring I/I from the private sector. Service lateral monitoring at the main line connection has not been commonly conducted because low flows and the presence of solids in a typical residential service lateral make collection of reliably accurate data extremely difficult. The depth measurements during dry conditions and wet conditions are used with water use records to calculate flowrates and volumes. Because lateral I/I is highly dependent on saturated ground conditions, service lateral I/I flow does not occur with every storm. It was found that when rainfall occurs during saturated ground conditions, service lateral I/I flows are a high percentage of the total I/I for the basin (see the Appendix, "WEF 2015 Private Property I/I Survey").

### 1.1.2 Estimating the Private Infiltration and Inflow Magnitude

Estimating private I/I flow levels in a system requires a combination of flow monitoring, system characteristic evaluation, and field testing. As previously mentioned, flow monitoring at the basin level can indicate whether the entire system has excessive I/I. A review of system characteristics typically follows to determine the likely source(s) of the I/I. If the basin has many building connections, then the private sector is potentially a significant contributor to I/I excessiveness. Also, data gathered by further investigation after flow monitoring may indicate that the area has many large private I/I sources such as area drains and downspouts. Balancing flow data results and I/I parameters to identified I/I sources is commonly conducted to estimate the private sector contribution to I/I. Unfortunately, limited quantification studies of private sector I/I contributions have been published (estimated contributions of private sector flow have been between 20 and 60% for a typical basin that has private connections). As stated, the survey provided in the Appendix of this publication and summarized in Table 2.1 ("Estimated

I/I Contribution Levels 2015 Survey") in Chapter 2, indicate that private sources activate at a higher percentage during wet years.

### 1.1.3   Establishing the Types of Private Infiltration and Inflow Sources

Flow monitoring at the basin level can provide insight to the types of I/I sources, including private I/I sources. If meter reaction to rainfall is consistent and has sharp peaks that quickly recede, then the sewer authority can expect directly connected I/I sources that may include outside private drains such as downspouts or driveway drains. If the meter reaction is only during saturated conditions, then directly connected outdoor sources are not likely and groundwater infiltration sources such as foundation drain connections and service lateral defects are likely.

## 1.2   Private Infiltration and Inflow Source Identification

Once flow monitoring has identified what parts of the system have excessive I/I, then source identification is typically conducted. The types of private I/I sources were previously presented in Section 2.0 of Chapter 2. Private I/I identification methods are described in this section and include typical methods that have stood the test of time and have been used for years in response to the Clean Water Act of 1972 (U.S. Congress, 1972). Also discussed are newer identification methods.

### 1.2.1   Smoke Testing

Smoke testing is a relatively low-cost method for identifying I/I sources. The primary I/I sources identified from smoke testing are inflow sources, which are sources that enter the system from rainfall at the surface. For the private sector I/I sources, smoke testing only identifies those sources located outside the building because the private building is not entered for this inspection.

Smoke testing is typically conducted by a three-person crew; one person operates the blower and the other two record data and location of sources. Smoke is introduced to the sanitary sewer system at manhole locations and forced through the pipe by a high-volume blower. The locations that could allow rainwater to enter the pipe are identified by the locations where smoke surfaces from the piping network, as illustrated in Figure 3.1.

Ground conditions must be dry for smoke testing to be effective. Wet ground conditions fill soil voids or cause soil to swell and can prohibit smoke from reaching the surface and locating the defect. Public notification before starting the smoke test is an important component of smoke testing procedures, especially for critical customers such as hospitals, emergency

**FIGURE 3.1**    Positive private property smoke source.

care centers, nursing homes, and schools, where activation of a fire alarm may cause disruption and possible injury. Local fire departments should also be included in the routine notifications.

### 1.2.2  Building Inspections

Building inspections are typically conducted by a two-person crew. During the building inspection, review of both outside and inside potential I/I sources is conducted. For the outside locations, the crew walks the property looking for suspect drains that could potentially connect to the sanitary sewer system, such as building downspouts piped underground with an unknown discharge location. For inside locations, internal building inspections are conducted mostly on the lowest floor of the building as inspectors look for stormwater connections to the sanitary sewer system. Typical building I/I sources are shown in Figure 2.3 in Chapter 2.

Older homes that tend to have groundwater seepage into the basement or crawl space location can often discharge to floor drains or sanitary sumps in a basement. During building inspections, a discussion with the resident about of past flooding experiences associated with wet weather can provide valuable insight to how stormwater could be entering the sanitary sewer system.

### 1.2.3 Dyed Water Testing

Dyed water testing is typically conducted to confirm I/I sources. Dyed water is applied to the source and if the dye ends up in the sanitary sewer system, connection of the source is confirmed. Also, dyed water testing can confirm if the source is directly or indirectly connected. When multiple I/I sources are suspected, a different-color dye can be used at each source to expedite testing and to verify each source. Dye testing can be conducted both preconstruction to confirm that a private I/I source is connected and postconstruction to verify that the source has been removed.

### 1.2.4 Service Lateral Closed-Circuit Television

In this chapter, the terms, "entire lateral", "service lateral", and "lateral", are used interchangeably, with each referring to both the upper and lower lateral as a whole. Investigating the service lateral pipe via closed-circuit television (CCTV) has been conducted for many years. but the intensity of service lateral CCTV has increased over the last 10 years. The increased use of this method is partially attributable to the improvements in technology that make lateral CCTV more practical and economical. Another reason for more CCTV of service lateral is the increased amount of directional drilling and boring of underground gas, water, and cable utility lines. These construction techniques have often damaged lateral pipes that cross easements for other utilities. Performing CCTV inspection of service laterals has, therefore, become increasingly necessary to determine if damage has occurred and to assess the severity of the damage.

Many inherent difficulties are encountered when televising laterals. One difficulty is accessing the lateral pipes, which may be complicated by owner resistance. Another complication may be definition of lateral ownership and the responsibility for subsequent repairs if defects are found. Another challenge is the smaller size of the service lateral pipes and how the lateral pipes may have been installed. Often, service laterals are found laid to inconsistent grades with bends (22.5- and 45-deg fittings) to match up with the house piping and then again with the main sewer piping. Private laterals are not routinely maintained like public sewer mains and are commonly found to have excessive accumulations of debris, roots, and structural damage. These problems can prevent collection of a clear and complete service lateral video, which makes it difficult to observe actual conditions inside the lateral.

#### 1.2.4.1 Cleanout Entry Closed-Circuit Television

Lateral inspections are commonly conducted from inside a house using a small push camera inserted to sewer cleanouts. This method is used by plumbers retained by residents to identify lateral issues that are causing backups in the building. Push cameras typically traverse less than 30.5 m (100 ft) from the house to the main. Often, the camera's position in the pipe is monitored by connecting sondes (i.e., a signal transmitter) to the camera and identifying the signal location with a surface receiver. When defects are identified, the sonde can be used to mark where the service lateral needs to be dug up for repairs.

### 1.2.4.2  From Main Sewer Closed-Circuit Television

Lateral inspections conducted from the main sewer use a launch system that has a lateral camera connected to the main sewer camera system. Main sewer lateral launch CCTV can span up to 30.5 m (100 ft) of the service lateral, which is the same limitation as that for the cleanout entry method. However, the advantage of the main sewer lateral launch CCTV is faster completion because the building does not need to be entered.

Sewer authorities with established sewer repair programs are focusing substantial effort on the lateral connection (i.e., the location where the lateral connects with the public sewer pipe). Use of main sewer lateral launch CCTV for pre- and post-rehabilitation of laterals ensures that the lateral connection is within the camera's reach. The main sewer lateral launch CCTV is often required when making repairs described in Section 2.3 of this chapter (entitled, "Private Lateral Repair Methods").

### 1.2.4.3  Electro-Scan Testing

A newer method of examining sewer pipes is electro-scan testing, which uses a probe to measure the electrical flow to and from the probe. Electricity flows through cracks and around corners, and can identify and locate defects that may not be visible to a CCTV investigator. The results are repeatable because the human element of visually identifying defects is removed. Detection is done solely by measuring electric current to and from the probe. Defects identified during scanning are located by linear feet from the start manhole, and the size of the defect is shown by peaks in current, which are displayed on a digital output screen.

For electro-scanning to work, the lateral pipe must not conduct electricity and water must be retained within the pipe. The scan is directly dependent on the amount of water within the pipe. If a complete circumferential scan is needed, then the section of pipe needs to be completely full. This can be accomplished by plugging one end of the pipe and filling the pipe with water for a short period of time. To avoid sewer backups into upstream services, full-pipe testing should be conducted during low-flow periods and with bypass pumping if needed.  More information regarding electro-scanning technology is available at http://www.electroscan.com.

## 1.3   Establishing Infiltration and Inflow Source Flowrates

When defining the means to contain peak wet weather flows within the system, improvements that remove I/I sources are considered an option, along with improvements that increase collection system and treatment capacity. Assigning I/I flowrates to private I/I sources is necessary to determine if the I/I source warrants removal. There are three methods commonly used to assign I/I flow to private sources: (1) the surface runoff estimate method, (2) the use of default flows for individual I/I source types, and (3) verification of flowrates through field testing. The method used typically depends on the source type.

### 1.3.1   Surface Runoff Method

The surface runoff method for calculating private I/I source flow is similar to the rational formula for storm sewer design. The process uses the equation

$$Q = CiA \qquad\qquad (3.1)$$

Where $C$ represents the surface runoff coefficient at a selected intensity ($i$) in millimeters per second (inches per hour) over the area ($A$) in hectares (acres). The typical sources for the surface runoff method are directly connected sources that are intended to drain surface area such as downspouts, area drains, and stairwell drains. These drains have been intentionally installed for removal of water from a protected area and discharging the water to a safe location.

### 1.3.2   Typical Private Infiltration and Inflow Source Rates (Default Flow Method)

The second method is the default flow method. A list of typical private I/I sources and a range of their typical flowrates in gallons per minute are provided in Table 3.1. These rates represent rates used on previous studies and a 1-year storm flow experienced in the Midwest for a 60-minute time of concentration. As an example, the downspout rate was estimated using the surface runoff equation shown in Section 1.31 (entitled, "Surface Runoff Method") to collect one-fourth of a typical roof ($A$ = 6.1 m (20 ft) × 6.1 m (20 ft), $i$ = 50.8 mm/hr (2 in./hr), $C$ = 0.9), which equals 28 L/min

TABLE 3.1   Private I/I inside and outside building sources.

| Defect type | Location | Flowrate, L/min |
|---|---|---|
| Downspout | Outside building | 22.7 to 37.9 (6 to 10 gpm) |
| Driveway drain | Outside building | 11.4 37.9 (3 to 10 gpm) |
| Service lateral general defects (cracks, breaks, offset joints, etc.) | Outside building | 0.4 to 1.9 (0.1 to 0.5 gpm) |
| Stairwell drain | Outside building | 22.7 to 37.9 (6 to 10 gpm) |
| Uncapped cleanout | Outside building | 0.4 to 3.8 (0.1 to 1 gpm) |
| Window well | Outside building | 1.9 to 7.6 (0.5 to 2 gpm) |
| Foundation drain | Outside building | 11.4 to 37.9 (3 to 10 gpm) |
| Area/yard drain | Outside building | 11.4 to 37.9 (3 to 10 gpm) |
| Sump pit | Inside building | 11.4 to 37.9 (3 to 10 gpm) |
| Sump pump | Inside building | 11.4 to 37.9 (3 to 10 gpm) |
| Basement drains | Inside building | 11.4 to 37.9 (3 to 10 gpm) |
| Service lateral—connection at main | Outside building | 1.9 to 11.4 (0.5 to 3 gpm) |

(7.4 gpm). The list also indicates which private I/I sources are found inside and outside of buildings. These flowrates should be adjusted to reflect basin-specific characteristics related to rainfall intensities for storm return intervals. Additional information related to I/I source flowrates can be accessed at https://www.rjn.com/blog/?s=flow+quantification.

### 1.3.3  Field Verification of Source Flows Method

The third method is to field-verify flowrates. There are two common types of testing to verify private I/I source rates: dyed water testing or measurement of flows from sump pumps using data loggers.

#### 1.3.3.1  Dyed Water Testing

Dyed water testing can be used to develop flowrates for actual sources by simulating rainfall entering the source. Capacity limitations of drains are dependent on its inlet control or pipe capacity. Knowing the flow being supplied to the source and the ability of the source to take the flow could indicate the maximum capacity of the source.

### 1.3.3.2  Data Loggers on Sump Pumps

A data logger can be added to one or more sump pump(s) that will represent flows for all sump pumps in a project area. Storm sump pumps that are connected to sanitary sewer plumbing inside buildings are a common private I/I source. A typical household sump pump is designed to pump 38 to 77 L/min (10 to 20 gpm) of flow. However, run times for sump pumps vary for storm events, antecedent moisture conditions, and subdrain system hydraulics. Also, sump pump discharges vary because of resulting friction losses of varying discharge-piping arrangements. Data loggers can measure pump "on" and "off" times that are used to estimate flows based on the rated capacity of the sump pump. Through a wet weather time span, an average flowrate can be determined.

## 1.4   Private Sector Infiltration and Inflow Source Data Management

### 1.4.1  Geographical Information System Utilization

A geographical information system (GIS) is a useful management tool to display selected data and to develop strategies for I/I source removal. Private sources of I/I identified during smoke testing, building inspections, and dyed water testing should have the following specific information about the defect recorded during the investigation:

- Type of defect,
- Address of the defect,
- Main sewer segment associated with the defect,
- Intensity of the defect's smoke,
- Surface area tributary to the defect,
- The GIS coordinates for the defect location, and
- Other characteristics applicable to the specific utility.

A GIS can include aerial photography or parcel information combined with data fields for each defect located by its GIS coordinates. This overlay of information spatially displays the data in relationship to buildings and streets. The GIS can print detailed maps that can be provided to property owners to begin the private I/I source removal process.

### *1.4.2 Asset Management Software*

Asset management software can be used as a repository for collected private I/I data. Most asset management software systems have built-in capabilities to work seamlessly with handheld digital data collectors and GIS mapping. Many of the newer systems are Web-based for instantaneous storage and availability of the collected data.

#### *1.4.2.1 Building Inspection Modules*

Building inspection modules of an asset management system typically provide a form that includes data fields for internal and external inspections of buildings. Each module also includes standard questions for building residents to allow the inspector to document useful information such as sewer backup and flooding history. Typical information stored in a building inspection module includes the following:

- Address,
- Property owner,
- Main sewer connection identification,
- Configuration of building structure (basement or no basement),
- Outside and inside I/I sources, and
- Photographs.

#### *1.4.2.2 Lateral Modules*

Lateral modules are used to help locate and provide property addresses and the material, size, and condition of laterals. Lateral conditions can be obtained through CCTV. Lateral CCTV observations from the main sewer are more prevalent since the advent of main sewer lateral launch CCTV equipment. Also, with pan and tilt capability in the main sewer cameras, main sewer CCTV can identify issues with the lateral connection to the main. The lateral connection condition is typically stored in the main sewer CCTV modules of asset management system software.

## 2.0 PRIVATE INFILTRATION AND INFLOW CORRECTIVE ACTIONS

Private I/I can be excessive and rob the capacity of collection systems. The private I/I flow can be combatted with preventive and corrective design methods.

## 2.1    Preventive Design Methods

Preventive measures are needed to reduce the effect that private-source I/I has on the economy, environment, and public health and to meet national and local regulations. The following subsections address new sewer design techniques that can be implemented to reduce the magnitude of private I/I sources.

### 2.1.1  New Sewer Design for Buildings

Reducing the likelihood of I/I entering the sanitary sewer system begins at the design stage of buildings that will be served by the systems. In lower-lying areas that experience frequent high groundwater and where homes require sump pumps, it is good practice to ensure that the sumps are deep enough that the collected groundwater remains below the invert elevation of the service lateral, essentially eliminating this path of potential infiltration.

To assist in preventing infiltration from entering through the lateral, a concrete ditch check can be installed in the service lateral trench. This ditch check is placed on the building side of, and a few feet from, the service connection at the main sewer. The ditch check works as a dam-type structure by preventing water from traveling along the service lateral trench to then infiltrate the sewer system through the service connection at the main sewer or a nearby defect in the main itself.

Internal building plumbing below the ground can also allow I/I to enter the system; to avoid this unwanted source, the plumbing system should be tested for watertight seals by air testing the system with compressed air. A successful test will maintain a constant pressure in the system for a minimum of 15 minutes (ASPE, 2012).

### 2.1.2  Private Connections to Existing Lines

While construction of the private connection is important, the inspection itself is equally as important. An inspector should only approve a connection that meets the project specifications. All service lateral connections to the main should be properly inspected before backfilling. Break-in connections are service lateral connection that are connected to the main by manually breaking a hole in the main line. Break-in connections install the service lateral connection at the break-in location using a saddle clamp to eliminate infiltration from entering the system. A break-in connection can be acceptable when made with clean cuts and watertight joints; however, it is not acceptable if a plumber attempts to shortcut the process by using an improper "hammer tap". The lack of inspection to ensure proper technique has led to significant quantities of infiltration entering many sewer systems at poorly made break-in connections.

A saddle connection, when properly installed, is effective for lateral connections. Instead of a hammer tap, a core-drilled hole with smooth circular edges should be made. The saddle connection is a polyvinyl chloride (PVC) hub with PVC or steel skirting that is the same curvature as the host pipe and is larger than the service lateral opening of the host pipe. A rubber gasket sits between the host pipe and the saddle connection, which is clamped and tightened to the host post with stainless steel clamps. Once the saddle is connected to the main sewer, the service lateral is then connected to the PVC hub on the saddle connections and is also held in place by stainless steel clamps.

Another method to eliminate infiltration via break-in connections is to use factory tees/wyes and fittings. Factory tees/wyes are generally 0.9 m (3 ft) in length and are connected to the main using couplers and lateral connection sleeves. The sleeve has an interlocking gasket that allows the service lateral to be pushed into the connection sleeve and locked into place. Using factory fittings allows the installers to have some rotational movement at the main line to properly align the service lateral to the factory fitting.

Another option to enhance elimination of I/I at a break-in connection is to connect the service lateral to the main using a PVC hub, a rubber sleeve, and a stainless steel band (e.g., an Inserta-Tee™, Universal Saddles, or ROMAC product; additional information related to these products is available at company-specific Web sites provided in the references section). A lateral repair made in this manner does not require any linear footage removal of the existing main. To further prevent infiltration, the break-in connection is made by core drilling the main line (i.e., creating a cleaner circular cut), installing the tee/wye, and connecting the service lateral to the tee/wye.

### 2.1.3  Design Considerations for the Effects of Best Management Practices on Laterals

Best management practices (BMPs) for stormwater management include features such as rain gardens and storm detention basins, which collect surface runoff and allow it to either permeate into the soils or be released slowly to creeks and other drainage ways. Plants and special soils included in the BMPs also provide treatment of storm runoff pollutants. Although these methods are attractive to storm sewer authorities, installation can cause I/I concerns for sanitary sewer authorities. A basic solution for sanitary sewer I/I is not to subject the system to rainwater. It is best to have stormwater directed to creeks by curb and gutter channels and to have hard (i.e., asphalt/concrete) surfaces and ground contours traveling away from the sanitary sewer. For sanitary sewers or laterals located under BMPs, exposure to the detained water allows more opportunity for the water to enter the collection system.

To combat the I/I effects of BMPs, sewer authorities are requiring developers to install BMPs away from the sewer system or to further protect the lateral from infiltration. For instance, where BMPs must be located over laterals, some sewer authorities are requiring encasements or other means to protect the lateral from being subjected to the detained water.

Best management practices can also be a positive for collection systems. Excess stormwater flow can be redirected from the combined system to the BMP, which effectively reduces the peak flows in the system. Also, rain barrels and rain gardens can be a positive green solution when removing private I/I sources such as downspouts or rerouting storm discharges away from the sanitary sewer system and to rain garden locations. Best management practices are also an excellent tool to help control stormwater flows and to provide natural treatment to waters; however, it is important to make sure that BMPs do not cause unintended consequences.

## 2.2    Corrective Design Methods

The following subsections describe methods to assist in removing private I/I from the collection system by making changes/repairs to existing I/I sources.

### 2.2.1   Cleanout Repairs

Defective cleanouts are typically identified during smoke testing. Repairing or replacing a broken or missing cleanout cap is an extremely affordable and efficient repair practice to effectively eliminate inflow at that location. Where conditions warrant that a plastic cleanout cap be protected from future damage, newer cleanout cap designs are available that include a small cast iron cover, as shown in Figure 3.2.

### 2.2.2   Downspout, Driveway Drain, and Area Drain Removals

Removal of private property downspout and area drain connections to the sanitary sewer are typically handled through a private-sector I/I removal program, as discussed in Chapter 4. Downspouts, driveway drains, and area drains are repiped to surface drainage areas. Depending on the contours around the drain, the source may need to be pumped to a storm sewer drainage way. In most cases, the underground downspout can be replaced with a spill gutter and splash block. Driveway drains and area drains are typically more complicated and require routing the drain several feet to a storm drainage way.

**FIGURE 3.2**   Post-cleanout installation and restoration.

### 2.2.3   Stairwell Drain Removals

Stairwell drains connected to the sanitary sewer system are often a difficult I/I source to remove. Most stairwell drains cannot be rerouted by gravity to the surface because of their relative depth below the surface. To reroute the flow often requires a pump system to send flow to the surface. This can be expensive when considering the typically small amount of I/I to be removed. Often, the lower-cost solution is to cover the stairwell with an awning to prevent rainfall from entering the drain.

### 2.2.4   Sump Pumps and Foundation Drain Removals

Sump pumps and foundation drain removals are the most intrusive types of I/I sources to remove. With these types of I/I sources, the property owner is required to make significant changes to the drainage system in the building or at the building footing outside. If the property owner has a storm sump pit, then the defect is much less expensive to correct. Adding a storm sump pump or redirecting the existing sump pump to the outside is generally the best solution. If no storm sump pump exists, then a new pit must be installed that requires considerable planning and design to connect with existing footing tiles. Identifying where the footing tiles connect to the service lateral may be the best way to locate where a storm sump pit needs to be installed. Special care needs to be taken when redirecting storm sump flows to outside the building to make sure that no building sanitary drain has been routed directly or indirectly to the storm sump pit.

Floor drains in basements and crawl spaces can collect water from basement wall seepage when the groundwater levels rise above the basement floor elevation. These conditions are more common in older homes with rock or stone basement construction. A common correction for these situations is to install a sump pit near the floor drain and pump flows outside of the building.

### 2.2.5  Redirecting Private Infiltration and Inflow Source Flows

When rerouting stormwater discharges, the designer should evaluate the surface conditions to minimize the negative effects of the new discharge location. For example, rerouting drains or pumped discharge from a newly installed storm sump pit to a poorly chosen location can cause drainage issues outside and, in some instances, icing at the discharge location. In some cases, the new discharge line may need to extend to the street or drainage way. Redirecting I/I source flows may also provide opportunities for green solutions. If agreeable to the property owner, rain barrels and/or rain gardens could be located at new storm discharge locations from downspouts and storm sump pumps.

## 2.3  Private Lateral Repair Methods

Repair of private lateral defects can eliminate a large amount of I/I. Property owners have a number of choices that can be used to repair laterals. If the lateral has multiple defects that affect the pipe's structural condition, then replacement of the lateral sewer may be necessary, either by open-cut or trenchless methods.

### 2.3.1  Complete Liner

A full-wrap, main-to-house, cured-in-place pipe (CIPP) lateral liner is a complete lateral rehabilitation approach. The main sewer segment at the lateral connection receives a full-wrap liner that covers the interior circumference of the main and is approximately 0.9 m (3 ft) centered at the lateral connection. A bladder is then inflated, pushing the CIPP liner through the lateral and against the interior of the lateral and main line, creating a new monolithic plastic pipe that extends from the main to the house.

### 2.3.2  Partial Liner

A short-connection full-wrap is the same as the full-wrap CIPP liner main-to-house; however, the length of the lateral liner is typically only 0.5 to 0.9 m (18 to 36 in.) in length. This is used when only the lateral-to-main connection is the focus of repair.

### 2.3.3  Grouting

Hydrophilic grout uses water as a catalyst and can have volumetric expansion of 5 to 8 times of that initially injected. Throughout the grout's curing stage, the surrounding soil becomes compacted and can create a flexible watertight seal; however, fluctuations in the water table can negatively affect the grout. As the water table decreases, the chemical grout, being a hydrophilic material it begins to lose moisture, which leads to shrinkage that allows I/I to enter the system. Chemical grouting is not an ideal option for structural rehabilitation; however, chemical grouting could be the appropriate choice for limiting the amount of I/I that enters the system.

Hydrophobic grout requires a catalyst that is not water and can expand up to 30 times the initial applied volume. Once cured, hydrophobic grout repels water, cures to a rigid state, and does not recover from compressive forces. Because of the rigidity, any movement from the structure will lead to cracks and breaks in the grout, allowing I/I to enter the system. As with hydrophilic grout, this is a viable choice if decreasing I/I is the ultimate goal, but grout is not a structural repair.

### 2.3.4  Replacement

Private laterals can contain offset joints, pipe-diameter changes, bends, and heavy roots, all of which can make CIPP lateral lining impractical and ineffective. In those instances, complete lateral replacement is often the best solution. The following subsections describe the open-cut and pipe-bursting approach to replacing service laterals.

#### 2.3.4.1  Open Cut

Open-cut replacement of laterals is often used as the most cost-effective repair. Open cut is generally conducted with a rubber-tired backhoe to minimize surface rutting during excavation, which is important when traversing on private property. A PVC replacement pipe is commonly used. Open-cut replacement can be quite intrusive and disruptive, especially for the construction of deeper laterals and in situations where surface conditions are expensive to replace such as driveways and sidewalks.

#### 2.3.4.2  Pipe Bursting

Pipe bursting is a trenchless rehabilitation technique that uses a pneumatic or static system to break and outwardly displace the host pipe and pull the new pipe into place. Both approaches require insertion and receiving pits. A pneumatic system uses a bursting hammer that is driven through the existing pipe by compressed air and a static system uses a constant tension winch in the receiving pit that pulls the breaking head through the existing pipe. This process is an option for lateral repair when trenchless replacement is favored over open cut.

## 2.4    Private Infiltration and Inflow Removal Programs

Private I/I removal programs consist of a systematic removal policy from identification of the private I/I source through confirmation of its removal. Examples of the various types of private I/I programs are included in Chapter 4 and more detailed case studies are included in Chapter 5. Utilities may choose to implement private I/I removal with in-house forces or with utility-paid contractors. Other utilities may choose to require the private property owner to implement private I/I removal. Property owners may be required to select contractors from a list of preapproved, licensed plumbers or contractors. If the property owner implements the private I/I corrective actions, most utilities require a follow-up inspection or testing to verify that the source has been removed.

Private I/I programs may include requirements for property owners to inspect their laterals and service connections. The inspection frequency varies with each program. Some sewer authorities require CCTV and repair of laterals during ownership transition, while others require lateral televising more frequently. Another system includes voluntary insurance coverages for private laterals. If defects are identified, then the insurance holder has their lateral repaired.

## 2.5    Public Information, Education, and Guidance Initiatives

Public information, education, and guidance initiatives on private I/I control are provided by organizations such as the Water Environment Federation (WEF). One source provided by WEF is the WEF Private Property Virtual Library (PPVL). The WEF PPVL is an "electronic library" where participating utilities with successful private property programs have volunteered to share examples of materials and lessons learned. The WEF PPVL can be accessed at http://www.wef.org. Information is also readily available from various other organizations such as the U.S. Environmental Protection Agency (U.S. EPA) and King County, Washington (at http://www.epa.org and http://www.kingcounty.gov/services/environment/wastewater/ii.aspx, respectively). Additional information and more in-depth discussions of private I/I removal and education programs and education and guidance initiatives are addressed in Chapter 4.

## 3.0  PRIVATE INFILTRATION AND INFLOW REMOVAL EFFECTIVENESS

Infiltration and inflow reduction can be measured by comparing preconstruction and postconstruction flow monitoring data. Another method is to track private I/I sources removed, assign a flowrate to the removed source, and total the volume of I/I flow removed from the system. The following sections discuss methods for evaluating the effectiveness of private sector I/I removal.

### 3.1  Preconstruction and Postconstruction Flow Monitoring

Preconstruction and postconstruction flow monitoring is often conducted on a basin level for I/I removal projects. Preconstruction and postconstruction flow monitoring is conducted when a significant number of sources are removed. A reduction in I/I flow must be achieved for a program to be considered successful. Control basins are often used to identify postconstruction storms that are suspected to have common storm reactions as preconstruction storms. The common parameters compared in preconstruction and postconstruction data sets are I/I rates and I/I volumes associated with rainstorms. Example flow monitoring data analyses can be found on the following city/county Web sites: http://www.kingcounty.gov, http://www.lakesuperiorstreams.org, http://www.3riverswetweather.org, and other links available from the King County, Washington, Web site.

### 3.2  Preconstruction and Postconstruction Private Infiltration and Inflow Source Testing

Preconstruction and postconstruction private I/I source testing can be conducted on an individual source basis. This may be a better testing method than basin flow monitoring because subtracting out public sector sources and varying antecedent ground conditions would not taint the results. Individual tests are typically conducted before and after construction with dyed water to verify that the source has been removed from the sanitary sewer system.

### 3.3  Private Infiltration and Inflow Removal Data Needs

Because of the inherent complications of identifying and removing private sector I/I sources, private I/I source removal has been deferred by sanitary sewer authorities. Because it has become evident throughout the years that private I/I sources have a significant effect on sewer systems and because this problem continues to compound when the private sewer system is ignored, new private sector I/I removal methods and approaches are continuously being developed. This trend will need to continue to bring excessive private sector I/I into manageable levels and to further eliminate sanitary sewer overflows to the nation's streams. Challenges that need further effort and solutions include how to better measure private I/I contribution and how to hold property owners accountable for their excessive I/I.

## 4.0   REFERENCES

American Society of Plumbing Engineers (2012) *Vent Systems, CUE 189*; American Society of Plumbing Engineers: Rosemont, Illinois.

Electro Scan, Inc. Technology Page. http://www.electroscan.com/ (accessed July 2015).

Inserta Tee Home Page. https://www.insertatee.com (accessed July 2015).

ROMAC Home Page. http://www.romac.com (accessed July 2015).

Universal Saddles. http://www.pipeconx.com/products/universal-saddles/ (accessed July 2015).

U.S. Congress (1972) Clean Water Act, 33. U.S.C. § 1251 *et seq.*

## 5.0   SUGGESTED READING

Bergi, V. (2014) What Does That Flow Look Like? RJN Group Blog Page. https://www.rjn.com/blog/?s=flow+quantification (accessed Sept 2015).

# 4

# Private Property Program Implementation Considerations

*Laurie Chase, P.E.*

| | | |
|---|---|---|
| 1.0 | STAKEHOLDER INVOLVEMENT | 42 |
| | 1.1 Customer Stakeholders | 44 |
| | 1.2 Internal Stakeholders | 44 |
| | 1.3 Local Plumbers, Service and Repair Contractors, and Builders | 44 |
| | 1.4 Real Estate Industry Interests | 45 |
| | 1.5 Regional and Multijurisdictional Programs | 45 |
| | 1.6 Other Stakeholders | 45 |
| 2.0 | PROGRAMMATIC ELEMENTS | 46 |
| | 2.1 Program Scope and Vision | 46 |
| | 2.2 Legal Authorities | 47 |
| | 2.2.1 *Sewer Use Ordinances* | 47 |
| | 2.2.2 *Construction Standards and Specifications* | 49 |
| | 2.3 Program Management and Staffing | 49 |
| | 2.3.1 *Management* | 49 |
| | 2.3.2 *Staffing Needs* | 50 |
| | 2.4 Public Education and Communication | 50 |
| | 2.5 Budgeting | 51 |
| | 2.6 Information Management | 51 |
| | 2.7 Standard Practices and Acceptable Technologies | 52 |
| | 2.8 Sustainability (Performance Metrics, Adaptive Management) | 52 |
| 3.0 | FUNDING CONSIDERATIONS | 52 |
| | 3.1 Expenditure of Public Funds on Private Property | 53 |
| | 3.2 Diversion of Funds from Other Customer Needs | 54 |
| | 3.3 Effects on Vulnerable Customers | 54 |
| | 3.4 Effects on Local Economy | 55 |
| | 3.5 Effects of Inaction | 55 |
| | 3.6 Utility Capital Improvement Funding Requirements | 55 |
| | 3.7 Funding Mechanisms | 56 |
| 4.0 | POLITICAL AND REGULATORY CONSIDERATIONS | 56 |
| | 4.1 Customer Equity and Environmental Justice | 56 |
| | 4.2 Private Property Rights | 57 |
| | 4.3 Local Political Issues | 58 |
| | 4.4 Federal, State, and Local Regulations and Enforcement | 58 |
| 5.0 | PRIVATE PROPERTY INFILTRATION AND INFLOW PROGRAM EXAMPLES | 58 |
| | 5.1 Enforcement-Based Program | 59 |

5.2  Point-of-Sale Lateral Inspection/Corrective
     Action Program                              60

5.3  Utility-Assumed Ownership and/or
     Operation and Maintenance of Privately
     Owned Lateral Program                       61

5.4  Publicly Owned (Lower) Lateral Focused
     Program                                     64

6.0  REFERENCES                                  65

Before implementing a private property program, all elements of the initiative should be carefully considered to encourage stakeholder buy-in, prevent unintended consequences, and foster continued success of the program. This chapter presents the following common elements of a successful private property program and how these elements can be developed within a stakeholder environment:

- Defined program vision, mission statement, goals, and scope;
- Legal authorities;
- Program management and staffing;
- Education and communication;
- Budget, funding, and procurement methods;
- Information management;
- Standard practices and acceptable technologies; and
- Sustainability (performance criteria and adaptive management approach).

It is important to note that utilities proceeding with private property infiltration and inflow (I/I) removal are not always motived by results of cost-effectiveness analyses. Sometimes, private property work is required for regulatory compliance (e.g., moratoriums, consent decrees, or other mandates), sewer separation work, or the integration of green infrastructure elements in an integrated planning scenario. Because of these varying motivations, one utility may need to focus more strongly on one programmatic aspect than another and/or may not need to fully develop all elements.

## 1.0  STAKEHOLDER INVOLVEMENT

This section describes common private property program-related stakeholders and the benefits of engaging a stakeholder group throughout program development and implementation. A stakeholder group, often referred to as a "public task force" or "special committee", may be created to provide an opportunity for meaningful public involvement during the creation and implementation of a sewer utility's private property initiatives. Stakeholders are individuals and organizations or their representatives who have an interest in the utility's work and policies or who seek to influence the utility's future direction. Public participation activities and processes allow the public to participate in the utility's actions and to hold the utility accountable for its decisions.

The U.S. Environmental Protection Agency (U.S. EPA) has published extensive guidelines for public stakeholder involvement. As part of the guidance documents, U.S. EPA's Office of Strategic Environmental Management maintains definitions of the most commonly used public stakeholder involvement terms at http://www.epa.gov/sites/production/files/2015-09/documents/stakeholder-involvement-public-participation-at-epa.pdf (U.S. EPA, 2001). This document defines "meaningful involvement" as follows:

- Potentially affected community residents have an appropriate opportunity to participate in decisions about a proposed activity that will affect their environment and/or health;
- The public's contribution can influence the utility's decision;
- Concerns of all participants involved will be considered in the decision-making process; and
- Decision-makers seek out and facilitate the involvement of those potentially affected.

In 2003, U.S. EPA published public involvement policy guidelines that describe key steps for various planning and stakeholder involvement options depending on the desired purpose and outcome (U.S. EPA, 2003). These stakeholder involvement activities incorporate a two-way transfer of information so that data, options, and outputs are provided and exchanged and advice and input is incorporated. The U.S. EPA guidelines provide information on how to develop a stakeholder action plan and include suggestions for building consensus between interested parties.

Key issues to be discussed within the stakeholder group setting include the program vision, mission statement, scope, and goal setting (both short and long term) and measures of success. Other issues that may be discussed include property owner/occupant notification, inspection methods, requirements for addressing identified defects (closed-circuit television [CCTV], smoke testing, dyed water testing, etc.), and possible repair, rehabilitation, or follow-up activities.

## 1.1    Customer Stakeholders

Depending on the type and scope of the private property initiative to be undertaken, some or all of the owners of different property types and uses may be stakeholders, including the following:

- Single-family residential,
- Multifamily residential (public and private housing),
- Commercial,
- Industrial, and
- Other.

The means by which property owners, residents, tenants, and customers are to be identified and included in the program and the level of authority and responsibilities each of these parties may have regarding sewer system construction, use (operation), maintenance, and repair should be discussed and agreed upon within the stakeholder group.

## 1.2    Internal Stakeholders

"Internal" stakeholders and the importance of each internal stakeholder's role in the success of a private property program should be recognized. Internal stakeholders may include the following groups:

- Governing officials (e.g., mayor and council [local government] or board of directors [special district]);
- Legal counsel;
- Accounting/finance staff;
- Public information officer;
- Engineering staff;
- Sewer maintenance staff;
- Inspection staff (i.e., I/I source investigation, building inspection, and sewer lateral replacement, rehabilitation, and/or repair inspection);
- Customer service staff;
- Public health officials; and
- City or local land use planner, forester, and arborist.

## 1.3    Local Plumbers, Service and Repair Contractors, and Builders

Organizations that build and/or service privately owned sewer laterals may play a key role in the success of a private property program and, therefore, should also be recognized. Representatives from these organizations could be invited to participate in the stakeholder group. Building good relationships with these stakeholders early on in the program planning stages will often result in smoother program implementation and a higher potential for building industry partnerships that will provide licensed plumbers who work within a building structure with licensed contractors who work outside of the structure.

For instance, one utility requested local plumbers and contractors to provide price quotes for distinct categories to be held for 2 years while I/I flows were separated from private residences. Other utilities have preapproved licensed plumbers and contractors to work on their programs.

## 1.4    Real Estate Industry Interests

Parties that have an interest in buying and selling properties are often affected by private property inspection requirements that are triggered at the time of a property's point of sale (POS). It is important to note that changes to inspection and corrective action requirements at the time of property transfer can affect the ability to complete real estate transactions. Therefore, it is vitally important to include representatives from these interested parties during the development of a POS program. Representatives of realtor groups and/or escrow officers may play an important role in ensuring the success of specific types of private property programs and should also be included in discussions that may affect property transactions.

## 1.5    Regional and Multijurisdictional Programs

When a regional wastewater authority or utility that serves multiple jurisdictions is developing a private property program that will affect more than one governmental entity, representatives from all potentially affected jurisdictions should be invited to participate in the stakeholder group. This consideration also applies when a county health department has jurisdiction on private property.

## 1.6    Other Stakeholders

Other stakeholders who are affected by wet weather overflows that result in contaminated surface water, groundwater, and ocean water should be identified. This may include water suppliers, recreational interests, the hospitality industry, fishery interests, and environmental advocates. These stakeholders may be affected by water supply issues and effects, beach closures, financial losses, or other issues, and should also be included in appropriate discussions.

## 2.0    PROGRAMMATIC ELEMENTS

A review of the case studies collected by the Water Environment Federation (WEF) Private Property Virtual Library (PPVL; accessed at www.wef.org) was conducted, and program elements that were common to the most successful programs were identified. These program elements include, but are not limited to, the following:

- Well-defined program scope and vision;
- Appropriate and necessary legal authorities;
- Established program management and adequate staffing allocation;
- Public education and communication;
- Adequate budget and program funding;
- Robust and accessible information management system;
- Written policies and procedures, standard practices, and approved methods and materials; and
- Continuous improvement approach with defined performance measures.

These elements are discussed further in this section.

### 2.1    Program Scope and Vision

Because of the challenging nature of private property-related issues faced by sewer utilities, most programs are implemented to address specific issues such as defective laterals that are found to be the root cause of sanitary sewer overflows (SSOs). Other focused private property programs may address issues such as maintaining access to easements so that preventive sewer cleaning and inspection can be conducted or educating homeowners about what not to flush down the drain.

As discussed previously, having a well-defined program scope and vision is critical to the success of any private property program and will enable a sewer utility to justify and support the resources necessary to implement the program. The scope should be re-evaluated as the program matures and, if necessary, changes to the vision and objectives may be made. For example, as the focus of the City of Santa Barbara, California's, Sewer Lateral Inspection Program moved from single-family home to multifamily residential properties over time, changes to the way in which the program was structured and implemented were necessary. Specifically, changes in the way property owners were identified and contacted were necessary in addition to ways in which the intent of the program was messaged and enforced.

## 2.2 Legal Authorities

A legal authority is adopted by elected officials and is enforceable in a court of law. This authority establishes the legal basis for the program, gives the sewer utility permission to conduct specified activities, and sets the legal boundaries of the customers' responsibilities. It is important that any changes to a wastewater utility's sewer use ordinance (SUO) or other legal authority are overseen by a municipal attorney and that if other jurisdictions are involved in a larger system that those jurisdictions' legal counsel are also given the opportunity to review proposed SUO changes before adoption.

Many communities rely on plumbing and building codes for the installation of sewer service and operation and maintenance (O&M) of the system. The following are two key drawbacks to using these codes to properly operate and maintain a sewer system:

1. Multiple codes are often referenced, and it is difficult to determine which codes are applicable to a sewer system, especially when dealing with private property issues.

2. These codes are typically adequate for the installation of the sewer service, to provide adequate funds for these activities, and to enforce that sewers are not properly operated and maintained. However, these codes typically lack language on operating and maintaining the sewers. A comprehensive approach used by communities to address these issues has been to create an SUO. The SUO may also serve to protect the environment by providing the legal mechanisms to prevent and mitigate SSOs and basement backups.

### 2.2.1 Sewer Use Ordinances

An SUO is used to describe a sewer utility's legal authority to control the quantity and type of flows discharged to its collection system and is written to address a utility's unique situation. Sewer use ordinances are used to regulate and provide standard remedies for common challenges, such as the following:

- Controlling or limiting quantity and quality of wastewater flows from new developments, satellite systems, and/or privately owned sewer systems;
- Identifying and requiring the disconnection of improper connections (including private sources of I/I);
- Accessing collection system components (including access to easements and testing for improper connections on private property);

- Requiring fats, oils, and grease source controls; and
- Requiring the proper design, construction, and inspection of new and rehabilitated sewers and sewer connections.

Sewer use ordinances often also contain pretreatment requirements if the utility is also the pretreatment control authority. In some cases, it may be useful for the utility to adopt a separate ordinance pertaining to private property sewer responsibilities. For example, the East Bay Municipal Utility District (Oakland, California) has separate wastewater control and private sewer lateral ordinances. In the past, these issues were combined in one ordinance, but were later separated into two ordinances for clarity purposes.

Utilities may consider strengthening the requirements for maintaining, repairing, and/or replacing the portion of the private sewer laterals that the utility is not obligated to maintain. Abilities to inspect and enforce the ordinance are critical. A clause could be included in the SUO that identifies the following:

- What constitutes a defective lateral or pipe or asset,
- The process for repair or replacement of a defective asset,
- The methods by which the defective pipes will be repaired or replaced,
- A schedule for the repair, and
- Consequences such as a specific penalty or fee or a lien for noncompliance.

Given a utility's liability for overflows is established under the Clean Water Act (CWA), the SUO could be revised to allow for enforceable I/I source corrections in private sewer systems. For example, the following provisions are contained in the City of Largo, Florida's, Ordinance 2005-01:

- Fact findings;
- Assignments of responsibility;
- Requirements for cleaning, inspection, and testing;
- Requirements for private system owners to maintain and repair sewer systems;
- Authorization for entry to test and inspect;
- Notice requirement before entering private property;
- List of defects that may impair system integrity;
- Standards for repair/replacement of defects;

- Time frame for cleanup of wastewater and repair/replacement of defects;
- Penalties for failure to comply, including civil and criminal; and
- Indemnification clause.

Sometimes, a wastewater utility will be directed to maintain private laterals or low-pressure systems to avoid SSOs or building backups, even though it is clearly stated in its SUO that the property owner is responsible for maintenance. It is recommended that the utility consider taking actions necessary to protect itself from unintended consequences by creating written policies, SUO amendments, or other legal authorities to account for these directives.

### 2.2.2  Construction Standards and Specifications

If not included in the SUO, a set of municipal construction standards, including details and specifications, should be adopted. All new sewer installation, rehabilitation, repair, and sewer connections should be inspected to ensure proper construction practices are followed. These standards will be helpful when subdivisions are initially started up, provided the guidance needed for properly installing all types of pipes and easement locations is available. Additionally, details and specifications for lateral repair and replacement and I/I source disconnections should be developed and adopted.

## 2.3    Program Management and Staffing

In a utility with a well-established organizational structure, staff and management devoted to implementing a private property program should be able to clearly articulate their job and position responsibilities. Additionally, all personnel should be trained to deal with potentially challenging people and situations and in conflict resolution.

A private property program's personnel requirement will vary in relation to the overall size and complexity of the collection system and the focus and scope of the program. In many cases, collection system personnel may be responsible for other duties. However, it is critical to the success of the program that an appropriate staffing plan be carefully considered and that adequate resources be assigned.

### 2.3.1  Management

The management strategy taken by a sewer agency can dictate workload. For example, a voluntary program would require less effort than a mandatory program with an enforcement component. Regardless of the size and scope of the program, it is important that the program is led by an administrator (i.e., a "champion") who can

- Promote the program and work with the public,
- Elicit buy-in and foster cooperation from all stakeholders,
- Appropriately budget for expected needs,
- Ensure that proper ongoing training for inspectors is provided,
- Measure the success of various program elements, and
- Clearly report progress to the local government and the public.

### 2.3.2  Staffing Needs

Private property programs are typically staffed with administrative personnel, on-site inspectors, and engineers or operations staff. Program personnel may also be augmented by staff from other departments, including a public information office, building department, or health department, or by contracted services such as CCTV inspectors or consulting engineers.

Some sewer utilities require that an experienced engineer with the authority to decide whether a lateral should be replaced or repaired is on-site during active mainline sewer rehabilitation projects. The field engineer is also given the authority to require drainage improvements on private property that are needed as inflow sources are removed from sanitary sewers.

## 2.4  Public Education and Communication

Many sewer utilities have stressed the importance of, and need for, public meetings to be held before, during, and after private property programs are established. Traditionally, utilities have provided public notification through Web sites, public television, radio, and local newspapers. Many utilities attend local festivals and public events, and may develop programs for elementary or vocational schools. More recently, utilities such as Dallas Water Utilities (Texas), DC Water (Washington, D.C.), and Seattle Public Utilities (Washington) have been engaging social media outlets such as Facebook and YouTube channels to provide meaningful, relevant information to their customers.

While the content included in the materials should be developed for targeted audiences and be supportive of the type and scope of program undertaken, information commonly provided includes a general introduction to the program; what is considered to be an illegal connection to the sewer system; what to do if wastewater enters a building; how to maintain backflow preventers, grinder pumps, ejector pumps, and other items; and how to properly care for a sewer lateral. Information regarding the consequences of not pursuing the program and the potential resultant environmental and financial risks can also help validate the program to the general public.

Public education materials may range from simple bill stuffers to comprehensive advertising campaign-type literature, glossy magazine advertisements, and cable television spots developed by in-house or contracted public relations staff. It is important to keep content posted on Web sites current so that the public is knowledgeable on program status, target areas, requirements, and results.

## 2.5    Budgeting

Identifying and allocating an adequate budget are critical to achieving a private property program's goals. Programmatic funding may be provided from a variety of sources, including user fees, property assessments, or appropriations from state or federal funding (see Section 3.7 of this chapter, "Funding Mechanisms", for more information).

A necessary element of budgeting is tracking program costs and using an annual baseline and accurate records each time the next period's budget is developed. The complexity and extent of a program's budget will vary by the type of initiative undertaken. Examples of items included in a private property program budget include labor, supplies, and contracted services.

Periodic reviews of the budget should be undertaken to consider if the program is properly and adequately funded in the context of the sewer utility's overall required initiatives.

## 2.6    Information Management

The management of data related to properties and property owners (including property type, contact information, inspection information, and contact dates/notes), collection system assets (pipes and manholes), and laterals (identified and geographically located) is an important program element. Some utilities use specific software programs for data asset management, recording budget-related items, and capturing performance metrics. Other utilities with smaller, or less complex programs may be able to record all program information in spreadsheets or databases.

Some utilities have captured sewer lateral locations in their geographic information systems. Parcel information may also be available from other departments/agencies. One sewer utility reported that local municipal building and zoning departments generally had plans for all buildings, and records of property storm flows were obtained from the local county drain commissioner.

An investment in a robust data management system may greatly reduce the staffing requirements of the programs utilizing the system.

### 2.7   Standard Practices and Acceptable Technologies

Sewer utilities should clearly define the applicable acceptable standard practices, methods, and materials for investigations, lateral repair and rehabilitation, new construction, and I/I source redirection/reconnection for their private property programs. Written policies, standard details, and guidance documents should be developed and made available to all affected parties for items such as source identification/condition assessment, rehabilitation/repair standards, certification, and inspection.

### 2.8   Sustainability (Performance Metrics, Adaptive Management)

To ensure sustainability, sewer utilities must focus on increasing overall business efficiency to maintain customer service and control costs. To do this, utility managers must understand existing service levels and the cost of doing business. The process of developing this understanding and improving efficiencies and effectiveness based on this understanding is called "performance management". Managing the performance of a private property initiative is an important consideration when creating the programmatic elements designed to achieve a sewer utility's overall private property-related goals. Key performance indicators (KPIs) that align with the utility's strategic plan and vision should be developed for all levels of the organization, and the proper information channels should be put into place that support the efficient data collection, reporting, and decision-making support.

The KPIs used to evaluate the progress of private property programs vary widely and should be specifically developed for each unique program. Examples of private property program-related KPIs used by sewer utilities include the following:

- Number of sewer laterals inspected per month,
- Percentage of sewer laterals with defects observed,
- Number of properties completing I/I source disconnections per month,
- Number of second notification letters sent to private property owners and/or residents per month, and
- Number and percentage of properties complying with POS requirements.

## 3.0   FUNDING CONSIDERATIONS

This section presents funding-related issues that are commonly encountered with private property-related programs.

## 3.1    Expenditure of Public Funds on Private Property

State statutes and administrative codes provide varying authority that cities or utility districts can use to demand corrective actions on private properties. Similarly, there are differences in the ability to use public funds for corrective actions on private property if those actions serve a cost-effective public purpose. In many states, utility districts have more flexibility in spending public funds on private property improvements than municipalities. Wastewater utilities in some states, including Tennessee, have successfully worked with legislators to modify state regulations related to the expenditure of funds for sewer lateral replacements. A wastewater utility should understand its legal options with respect to private property corrective actions.

Any time a utility spends money to correct or improve private property assets, questions will arise regarding the legality and equitability of such public investments. While determinations must be made on a utility-specific basis, it is important to consider whether existing local or state laws restrict use of public funds on private property or the use of utility rates to disproportionately benefit certain customers, such as California's Proposition 218. California voters passed Proposition 218 in 1996, which added Article XIIID, Section 6 to the California Constitution. Article XIIID, Section 6 contains three substantive provisions: (1) the revenue from fees and charges shall not exceed the cost to provide service; (2) the revenue from fees and charges must not be used for any other purpose; and (3) fees and charges shall not exceed the "proportional cost of service attributable to the parcel." Article XIIID does not expressly prohibit the use of public funds for private uses. For example, it is common for water utilities in California to provide funding to their customers to remove turf to reduce demand; providing this funding has thus far been considered compliant with Article XIIID. Providing funding for private lateral retrofits is similar. In either case, the substantive provisions of Article XIIID must be met.

Some utilities have sought a determination from the state attorney general before embarking on a private property I/I removal program. If, from a legal standpoint, private owners will derive some benefit from the program (provided that it is incidental to the public benefits), utilities may still need to consider the political implications of any perceived fairness issues among customers if funding or resources are made available to some, but not all. It should be noted that even in California, where Proposition 218 is in effect, utilities have still implemented programs that fund private property work as those programs have typically been justified based on the overarching public good.

Additionally, utilities have sought guidance from the Internal Revenue Service (IRS) or tax attorneys as to whether grants or other uses of public funds on private properties could be considered taxable. Recent responses to IRS inquiries indicate that determinations must be made on a case-by-case basis.

## 3.2   Diversion of Funds from Other Customer Needs

When discussing funding sources with stakeholders and customers in public forums, many utilities outline the potential effects of allocating public funds for private property-related programs. These effects may include fewer available funds for regulatory-required wet weather remediation projects (e.g., sewer capacity augmentation, retention basins, and water resource recovery facility [WRRF] upgrades), repair and replacement of defective sewer lines, and other improvements necessary to continue to provide expected levels of service to the entire customer base. By delaying the investment in these types of initiatives, higher capital budgets and increased sewer rates may be required in the future.

Sewer utilities may also consider discussing the alternatives to private property programs and how much those would cost with stakeholders and customers. For example, a private property program may cost $5 million to implement, whereas the construction of storage facilities that would be needed if the private property-related I/I were not removed would be $20 million. It is important that context is provided not just for what the private property program takes away from, but also where it provides financial benefit. This adds to the argument that private property work benefits the greater service area and is not a gift of public funds to a specific group of private property owners.

## 3.3   Effects on Vulnerable Customers

The financial effect of addressing public and private sources of I/I can create difficult challenges for local governments. In many cases, private property program costs will have a disproportionate burden on lower- and fixed-income households.

U.S. EPA Combined Sewer Overflow Program Guidance adopts the approach used for assessing substantial and widespread economic and social effects of water pollution control projects under 40 CFR 131.10(g)(6). Although this guidance may not be directly applicable for publicly funded private property programs, the fundamental principle of a financial capability assessment is based on the understanding that the ability of a wastewater utility to meet CWA objectives while remaining sustainable and affordable rests on the financial capability of individual households. These households make up the community and, collectively, are the owners of public utilities and are responsible for the fiscal support of all public services and public infrastructure through rates, fees, and taxes.

As mentioned previously, many wastewater utilities have been unable to implement assistance programs because of prohibitions against using public funds on private property. A wastewater utility should consider the effects on vulnerable customers before embarking on a program that is not mandated by a regulatory requirement.

In cases where a sewer utility is facing regulatory requirements, it is important that any affordability assessment consider the cost of any private property program to property owners, in addition to the work the utility may be funding. For example, in assessing the effect to each household, the private property burden should be added to the rate effect of any utility work.

## 3.4    Effects on Local Economy

Some local businesses may benefit from private property programs, including licensed plumbers, rehabilitation contractors, and landscapers. Additionally, businesses, commercial establishments, and industries that have been affected by sewer capacity issues may realize reduced backups and fewer claims. Conversely, some entities may go out of business or move out of the service area if the I/I-related problems are not addressed or if the cost of private property programs on business owners is too high.

## 3.5    Effects of Inaction

When private sources of I/I are not adequately addressed when they are responsible for resultant sewer capacity exceedances, SSOs, combined sewer system releases, building backups, regulatory fines, and poor community relationships may result. Customers who have experienced sewer backups will continue to have problems and property values may reduce. In addition, future budget increases may be difficult to get passed. A wastewater utility may be forced to spend public monies on expanding facilities to handle I/I flows instead of upgrading existing equipment or expanding the public sewer system.

However, it should be noted that, in some cases, the cost of a private property program (on the utility and/or the property owners) might be higher than the cost of inaction.

## 3.6    Utility Capital Improvement Funding Requirements

It is important to note that many funding agencies may require a cost-effectiveness analysis to authorize collection system work when granting or loaning funds to wastewater utilities. In recent cases, the U.S. Department of Agriculture Rural Development and U.S. EPA State Revolving Fund programs have acknowledged that private property issues affect the overall system and will fund cost-effective solutions. Ultimately, I/I is a precursor to more catastrophic problems, including pipe failures and SSOs, and, as such, public and private sources can be incorporated to the cost-effective analysis.

## 3.7   Funding Mechanisms

The type of mechanism that can be used to fund private property programs is varied and may be limited by local, state, and federal regulations and requirements. A wastewater utility should thoroughly investigate methods by which its program can be initially funded and sustained over time.

Potential funding sources for private property programs include, but are not limited to, the following:

- Utility pays entire cost of the private property initiative;
- Utility subsidizes the program through capped costs paid by property owners;
- Utility provides partial grants and/or works with federal, state, and/or local funding agencies to provide property owners with partial grants;
- Utility establishes a grant program for vulnerable customers;
- Utility establishes a zero-interest loan program;
- Utility participates in renewal insurance programs (e.g., Sandy Suburban Improvement District, Sandy, Utah) and property owner pays for insurance policy premium;
- Property owner pays for entire cost of lateral repair or replacement and/or I/I source mitigation; and
- Special levies (promote maintenance).

# 4.0   POLITICAL AND REGULATORY CONSIDERATIONS

This section will present political and regulatory considerations that have not yet been discussed.

## 4.1   Customer Equity and Environmental Justice

Environmental justice, as defined by U.S. EPA, is the "fair treatment and meaningful involvement of all people regardless of race, color, national origin, or income with respect to the development, implementation, and enforcement of environmental laws, regulations, and policies" (U.S. EPA, 2015). Environmental justice will be achieved when everyone enjoys the same degree of protection from environmental and health hazards and equal access to the decision-making process to have a healthy environment in which to live, learn, and work. Customer equity and environmental justice is particularly important in communities where portions of the community include low-income, disadvantaged populations.

Considerations should be made for customer equity and environmental justice when deciding whether a private property program will be voluntary or mandatory, how it will be funded, and, if necessary, how enforcement will be accomplished. Appropriate methods of program information dissemination should be undertaken for all customers, and consistent and fair notification and investigation techniques should be applied for owners and residents of each property type.

Issues such as the following should be discussed openly with stakeholders:

- What parts of the system to focus on first—for instance, areas that are experiencing building backups or that contribute the highest rates of I/I, regardless of income levels and ZIP code?
- Will single-family residences, multifamily residences, and housing complexes be included in the program?
- Will businesses and commercial establishments be included in the program, and how will they be treated from a financial and enforcement standpoint?
- How will property owned or operated by the federal government or state or local entity be addressed?

## 4.2   Private Property Rights

Private property rights must be maintained, but SUO language needs to clarify the rights of the sewer utility to access private property and/or redefine what infrastructure is owned by the private property owner and which is owned by the utility.

Specific property rights should be evaluated regarding vacuum and low-pressure pump systems (e.g., grinder pumps and septic tank effluent pumps). These systems have been used in areas with shallow bedrock and high groundwater tables such as lakes and other low-lying areas, and are sometimes promoted as a possible way to address private sources of I/I. The issue of who is responsible for maintaining the infrastructure must be clearly identified in the wastewater utility's SUO. Some utilities have specific language included in their sewer easements to address pressure sewer system initial construction, perpetual easement, significant repair easement, easement restrictions, rights of restoration, and necessary damages release (e.g., Licking County Water and Wastewater Department's Buckeye Lake District No. 1 Sewer Easements [Ohio]).

### 4.3  Local Political Issues

High-profile wastewater utility initiatives such as private property programs can be influenced heavily by local political issues, especially during election campaigns; when dealing with significant businesses or industries; and when utility managers are forced to address specific interest group, council member, board member, or other political representative's concerns, even though property owner responsibilities are clearly stated. For these reasons, it is important to include influential leaders and members of political bodies in the stakeholder process.

### 4.4  Federal, State, and Local Regulations and Enforcement

A utility's motivation for creating a private property program may be a result of a CWA violation, a Combined Sewer Overflow Long Term Control strategy, a National Pollution Discharge Elimination System (NPDES) requirement, a state-mandated Capacity Management Operation and Maintenance Program requirement, or other regulatory action.

It is important that the wastewater utility clearly defines all regulatory drivers for its private property initiatives, including the ramifications of not achieving compliance (e.g., not reducing I/I to acceptable levels if that is an NPDES permit requirement). Additionally, the utility should clearly identify all parties, including satellite systems and large industrial and institutional customers, that may play a role in achieving compliance and include them in the stakeholder engagement process. The importance of this issue is demonstrated further in the example private property program described in Chapter 5, Section 4.0, Case Study: East Bay Municipal Utility District.

## 5.0  PRIVATE PROPERTY INFILTRATION AND INFLOW PROGRAM EXAMPLES

The WEF PPVL contains information regarding a number of private property programs implemented throughout the United States. Some of these programs can be categorized into the following general types:

- Enforcement-based programs,
- Point-of-sale lateral inspection/corrective action programs,

- Utility-assumed ownership and/or O&M of private lateral programs, and

- Publicly owned (lower) lateral focused programs.

The following subsections provide a WEF PPVL example for each of the aforementioned types of programs. More details on each of the examples can be found at www.wef.org/privateproperty.

## 5.1    Enforcement-Based Program

Enforcement-based private property type programs are typically implemented in response to a regulatory directive placed on a sewer utility (e.g., consent decree or administrative order). As part of this type of program, a local sewer utility will investigate sources of I/I on targeted private properties and determine that I/I removal corrective actions are needed. The utility will then require the property owner to undertake, and to pay for, the required actions. An example of this type of program was undertaken by the City of McMinnville, Oregon.

McMinnville is a city of approximately 30,000 people located midway between the Pacific coast and Portland, Oregon, about 48 km (30 miles) from the capital city of Salem, Oregon. McMinnville's wastewater collection system totals about 1448 km (900 miles) of sewers, with about 1% comprising combined sewers and the rest separate sanitary sewers. The city's WRRF discharges treated effluent to the South Yamhill River.

The city was directed by U.S. EPA and the Oregon Department of Environmental Quality to control the overflow of untreated wastewater into the Yamhill River. These overflows occurred in part because the city's collection system was severely deteriorated in some areas. When the current plant was evaluated in 1996, the city estimated that approximately 60% of the I/I entered the system from private sewer laterals, which, in McMinnville, is defined as the upper lateral from the edge of the street right-of-way to the building.

In 1997, the city implemented a private sewer lateral replacement program as part of its I/I control program. This program is still in place and requires property owners to replace the privately owned portion of the lateral line when the city deems the lateral defective. Property owners are informed of their responsibility by a letter from the city and given 90 days to repair or replace the private lateral. A rebate incentive of 10% of the cost, up to $250, is offered to encourage property owners to complete the work within the 90 days. If the work is not done in 90 days, the second phase of the enforcement program begins. Under the second phase, the property owner is allowed 10 months to finish the work, but is assessed a $50-per-month penalty plus interest. The penalty is waived if the work is completed within the 10 months. Property owners in noncompliance continue to accumulate penalties at the rate of $50 per month plus interest, and a lien is placed on the property.

Lateral inspections are conducted when the city rehabilitates or replaces sewer mains and the publicly owned lower lateral or when a property owner or tenant reports a lateral problem to the city. As result of a lateral inspection, a condition grade is assigned to each lateral. The lateral age, material, and condition grade are then used to decide whether or not the privately owned upper lateral needs to be replaced. Typically, the city also replaces the public portion of the lower lateral and installs a cleanout at the right-of-way at the same time as the private replacement of the upper lateral. The cost of the lower lateral replacement and cleanout installation is typically between $1,500 and $2,500.

In 2015, the city estimated that approximately 1875 laterals were inspected since the program was implemented in 1997, which corresponds to approximately 18% of the laterals in the service area. Between 30 and 40% of the laterals passed the inspection and did not require replacement.

## 5.2    Point-of-Sale Lateral Inspection/Corrective Action Program

Point-of-sale programs typically require property owners to inspect private laterals when the property ownership is transferred. If private lateral defects are identified, the property owner is required to implement corrective action to repair, rehabilitate, or replace the lateral. An example POS program operated by the City of San Bruno, California, is summarized here.

Effective May 8, 2015, the City of San Bruno began enforcement of a POS requirement for sewer lateral inspections. The San Bruno City Council approved the requirement April 8, 2014, with implementation effective May 8, 2015 (Municipal Code Chapter 10.13; Ordinance No. 1822). As with other POS mandates for sewer lateral inspections in San Mateo County, this is a result of a judicial consent decree. The code requires property owners to obtain a Sewer Lateral Compliance Certificate (hereafter referred to as "compliance certificate") when a residential property was originally constructed 50 years or more from the date of recordation of a deed or mortgage.

Before the deed transferring the property is transferred to or vested in any other person or entity, the property owner is required to conduct an inspection of the sanitary sewer lateral at his or her own expense. A copy of the video inspection must be given to the Public Services Department for review. While the city does not require a specific company conduct the inspection, certain requirements must be met for the video to be acceptable. Any subsequent repair or replacement work deemed necessary as a result of that inspection must be completed and approved by the city before issuance of a compliance certificate and transfer of title. The following properties are exempted from the requirement:

- Condominium or cooperative apartment buildings or the units within those buildings, except as a condition to conversion to a condominium or cooperative apartment building;
- Properties that share a sanitary sewer lateral with another property;
- Property transfers that do not involve the payment of a county transfer tax;
- Properties that have been issued a Certificate of Compliance by the city within the past 5 years;
- Properties that received an acceptance of a test pursuant to the POS program's requirements within the past 5 years, if partial or no repairs of the lateral were required and any repairs were completed pursuant to permit and inspection by the city;
- Laterals that were repaired or relocated, inspected, and approved under the city's program within the past 5 years; and
- Properties for which the city provided an acceptance of construction work for a sewer lateral replacement within the past 20 years.

The process for obtaining a compliance certificate is handled by the Public Services Department and is depicted in the flow chart shown in Figure 4.1. A Time Extension Certificate may be considered by the city when a compliance certificate cannot be obtained before title transfer. To be eligible for this extension, the property owner must provide a written request to the public services director for a time extension of up to 180 days, in which time the inspection and/or repairs or replacement required by the code must be performed.

## 5.3    Utility-Assumed Ownership and/or Operation and Maintenance of Privately Owned Lateral Program

At times, a utility may assume ownership and/or O&M responsibility for both the lower and upper laterals. This can be beneficial for both the property owner and the utility by, respectively, having professional collection system operators responsible for the unseen underground pipes and facilities and by improving customer relations particularly in cases where a utility crew has already arrived on-site to investigate the cause of the customer's problem.

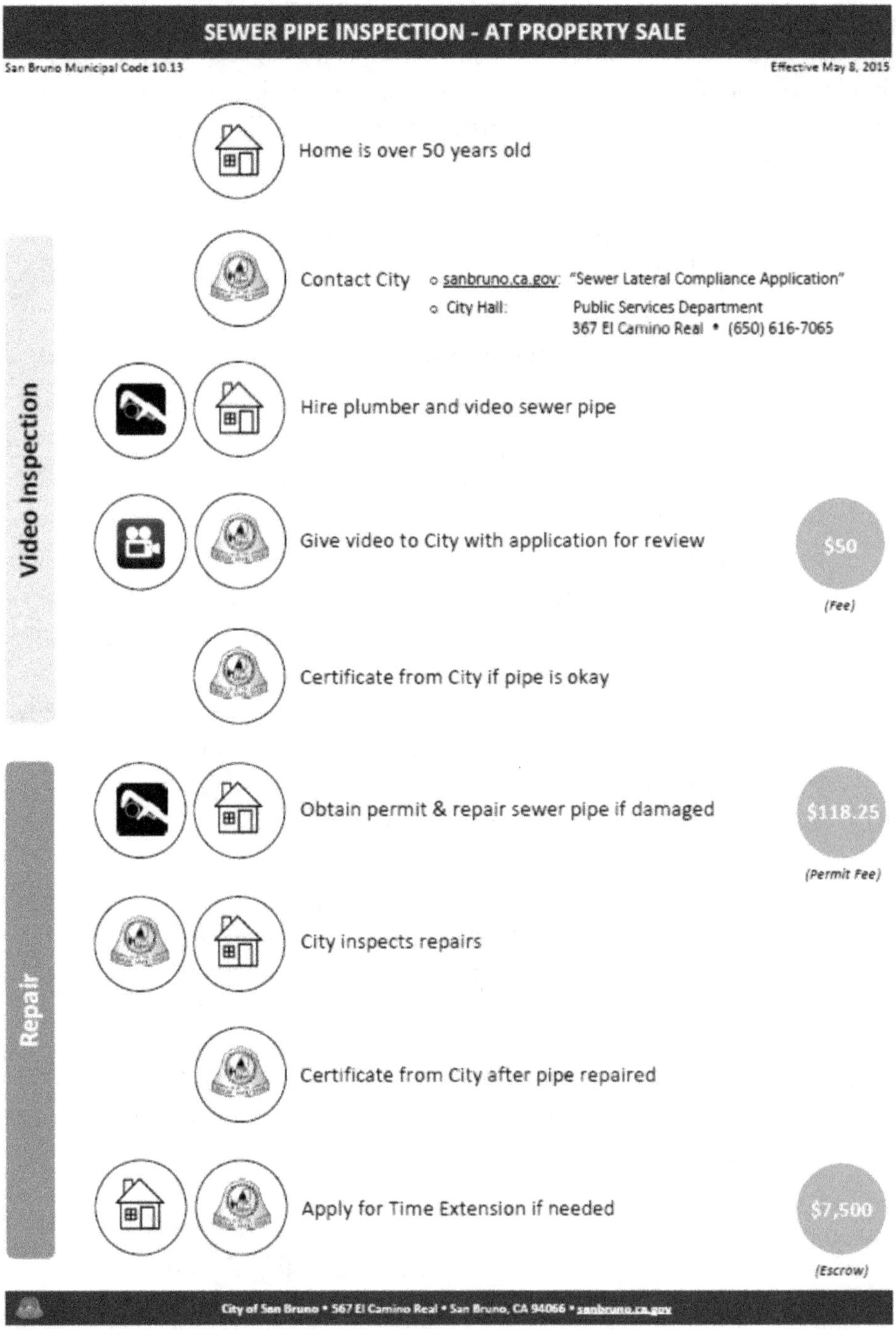


**FIGURE 4.1** City of San Bruno POS flow chart.

The Grand Strand Water and Sewer Authority (GSWSA), based in Conway, South Carolina, is a regional water and wastewater utility founded in the early 1970s; as of 2015, it serves more than 73,000 customers. The GSWSA provides services from the North Carolina line south to parts of Georgetown, Dillon, Marion, and Horry counties, in addition to rural areas west of the Waccamaw River. The GSWSA provides contractual services to six of the county's municipal water and sewer utilities, including the coastal communities of Myrtle Beach and surrounding cities, and two public water companies.

The GSWSA has historically been responsible for the lower lateral and the street cleanout, with the property owner responsible for the upper lateral, house cleanout, and house plumbing. An evaluation of sewer complaints made during 2001 indicated that GSWSA responded to a total of 566 sewer backup complaints, 58% of which were determined to be GSWSA's responsibility. However, GSWSA staff expended the same response effort for 42% of complaints that were not its responsibility. Further, many of the customers informed of their responsibility to resolve the problem themselves did not understand why GSWSA crews could not resolve the problem immediately.

The GSWSA managers and staff evaluated how to better manage, operate, and maintain its infrastructure, and how to provide improved customer service. As a result, GSWSA determined that it could cost-effectively provide O&M services for both the lower and upper laterals by implementing a Service Line Maintenance Program. Under this program, GSWSA

- Maintains the water and sewer lines from the main line to the exterior of the house;
- Repairs damages incurred during repairs to water or sewer lines, including repair of sidewalks, driveways, or landscaping damaged in the course of the repairs; and
- Provides reimbursement for damage caused by repairs.

New sewer customers are automatically enrolled in the program, while existing customers may request the service. Customers that fail to pay the monthly charges do not receive the service line maintenance services. Customers may discontinue the service, but, if a repair has been made by GSWSA, the Service Line Maintenance Program on the customer's account must be active for a minimum of 3 years before cancellation. Customers are encouraged to call GSWSA before calling a plumber because GSWSA may be able to resolve the problem for free.

Following implementation of the program, repeat sewer backups have been significantly reduced. At the end of June 2015, GSWSA had 49,018 sewer customers active in the program, which is approximately 67% of the total sewer customers. The sewer revenue received under this program has more than covered the additional costs incurred by GSWSA.

## 5.4    Publicly Owned (Lower) Lateral Focused Program

Based on a 2015 WEF Private Property I/I Survey (included as an Appendix), approximately 47% of the utility respondents accept ownership and O&M responsibility for the lower lateral (as defined by the utility). One utility that has implemented a program focused only on the publicly owned portion of sewer laterals is the San Antonio Water System (SAWS) in San Antonio, Texas.

The City of San Antonio created SAWS as a new entity on May 19, 1992. The SAWS operates more than 16 900 km (10 500 miles) of water and sewer mains serving a 1663-km$^2$ (642-sq mile) area; it also operates an extensive reclaimed water system consisting of 203 km (126 miles) of mains. The utility is responsible for sewer mains, lower laterals, and related appurtenances. The property owners are responsible for the upper lateral and building plumbing.

The SAWS offers a Sewer Lateral Reimbursement Program for eligible customers who follow these instructions:

- Call SAWS to report a sewer lateral blockage (e.g., a toilet or tub is backing up). The SAWS will immediately dispatch a crew to check the sewer main.
  - If the sewer main is clogged, SAWS will clear the blockage.
  - If the sewer main is clear, it is assumed that the problem probably lies between the house and the sewer main, and the customer is advised to call a plumber.
- Call a licensed plumber with electronic locating equipment.
  - If the licensed plumber finds the problem within the property line (i.e., the private upper lateral), the property owner is responsible for directing the licensed plumber to fix the problem. The SAWS will not reimburse the customer for the plumber's time or service.
  - If the licensed plumber finds that the problem is beyond the property line (i.e., the lower lateral in the right-of-way, street, or alley), the plumber must electronically locate the lateral, mark the location of the obstruction with green paint, and contact the SAWS Emergency Services Section. The SAWS will promptly respond and make the necessary repairs.
- If SAWS crews make necessary repairs, customers mail the licensed plumber's receipt for normal and customary charges to SAWS for reimbursement within 4 to 6 weeks.

Plumbers are instructed to provide an estimate to the customer before beginning any work. Further, the plumbers are notified that SAWS only reimburses the customer based on normal and customary plumbing charges related to electronically locating the problem. Reimbursements are not provided for property improvements, such as new cleanouts. If excavation is necessary and the break is not located, SAWS will contact the plumber at that time. Plumbers who fail to respond or properly locate the break are subject to a time and materials charge.

## 6.0    REFERENCES

U.S. Environmental Protection Agency (2015) Environmental Justice. http://www.epa.gov/environmentaljustice/ (accessed Sept 2015).

U.S. Environmental Protection Agency (2003) *Introducing EPA's Public Involvement Policy.* http://nepis.epa.gov/Exe/ZyPDF.cgi/100045RR.PDF?Dockey=100045RR.PDF (accessed Feb 2016).

U.S. Environmental Protection Agency (2001) *Stakeholder Involvement & Public Participation at the U.S.EPA Lessons Learned, Barriers, & Innovative Approaches.* http://www.epa.gov/sites/production/files/2015-09/documents/stakeholder-involvement-public-participation-at-epa.pdf (accessed Feb 2016).

# Private Property Infiltration and Inflow Program Case Studies

*Aaron Witt, P.E.; Christopher Ramsey, P.E.;*
*Richard (Rick) E. Nelson, P.E.; and Dan Ott, P.E.*

| | | |
|---|---|---|
| 1.0 | WATER ENVIRONMENT FEDERATION PRIVATE PROPERTY VIRTUAL LIBRARY | 68 |
| 2.0 | JOHNSON COUNTY WASTEWATER, KANSAS | 69 |
| | 2.1 Utility Background | 69 |
| | 2.2 Private Infiltration and Inflow Removal Program | 70 |
| |     *2.2.1 Program Drivers* | 70 |
| |     *2.2.2 Program Characteristics* | 71 |
| |     *2.2.3 Public Outreach* | 72 |
| |     *2.2.4 Program Resources and Tools for Source Identification and Removal* | 73 |
| |     *2.2.5 Program Costs* | 74 |
| | 2.3 Program Effectiveness | 74 |
| | 2.4 Conclusions | 77 |
| 3.0 | KING COUNTY, WASHINGTON | 79 |
| | 3.1 Utility Background | 79 |
| | 3.2 Private Infiltration and Inflow Removal Program | 80 |
| |     *3.2.1 Program Drivers* | 80 |
| |     *3.2.2 Program Characteristics* | 80 |
| |     *3.2.3 Public Outreach* | 80 |
| |     *3.2.4 Source Identification and Removal* | 81 |
| |     *3.2.5 Program Resources and Tools* | 81 |
| |     *3.2.6 Program Costs* | 82 |
| | 3.3 Program Effectiveness | 82 |
| | 3.4 Conclusions | 83 |
| 4.0 | EAST BAY MUNICIPAL UTILITY DISTRICT | 84 |
| | 4.1 Utility Background | 84 |
| | 4.2 Private Infiltration and Inflow Removal Program | 84 |
| |     *4.2.1 Program Drivers* | 84 |
| |     *4.2.2 Program Characteristics* | 85 |
| | 4.3 Program Effectiveness and Conclusions | 88 |
| 5.0 | SUMMARY | 88 |
| 6.0 | REFERENCES | 88 |

The previous chapters of this publication demonstrated many approaches and solutions that can be taken to remove infiltration an inflow (I/I). This chapter presents case studies of three utilities and gives an overview of the drivers to implement a private I/I program. Program approach, resources, costs, effectiveness of programs, and lessons learned are also discussed. The case studies were selected based on different program approaches, questionnaire responses, and using information available in Water Environment Federation's (WEF's) Private Property Virtual Library database (PPVL; accessed at www.wef.org). An overview of the WEF PPVL is also discussed as a resource containing information on additional utilities and their approach to removing I/I.

## 1.0  WATER ENVIRONMENT FEDERATION PRIVATE PROPERTY VIRTUAL LIBRARY

WEF's PPVL is a growing library of case studies from private property-related programs at wastewater utilities. It was developed by a WEF Collection System Committee (CSC) Project Team and is intended to be a resource for utilities seeking information or advice about private property programs. The WEF PPVL is designed to be a resource for utilities that are

- Evaluating the options and benefits related to different types of private property programs, such as those examples described in Section 5.0 of Chapter 4;
- Developing a private property program to implement in their service area; and
- Enhancing existing private property programs.

The library includes information gathered from successful private property programs targeting

- Sanitary lateral repair or replacement,
- Infiltration and inflow source detection and elimination,
- Lateral condition assessment,
- Privately owned pumping station operation and maintenance (O&M), and
- Sewer easements.

Detailed information includes resources such as

- Utility completed questionnaires,
- Public education materials,
- Outreach letters to residents,
- Ordinances and codes,
- Operating and maintenance procedures,
- Inspection forms, and
- Design details and specifications.

The project materials and forms used by the case studies presented in this chapter can be found in the WEF PPVL at www.wef.org.

The following sections of this chapter present case studies of successful private property I/I programs at Johnson County Wastewater (Kansas); King County (Washington); and East Bay Municipal Utility District (California).

# 2.0  JOHNSON COUNTY WASTEWATER, KANSAS

Johnson County Wastewater (JCW) has achieved high levels of collection system performance, including reduction of sewer overflows and basement backups, by improving system O&M practices, completing extensive public sector sewer rehabilitation, and managing a private sector I/I source/defect removal program since the 1980s.

## 2.1  Utility Background

Johnson County Wastewater provides sanitary sewer service to more than 500,000 customers in Johnson County, Kansas. Its service area covers 421.2 km$^2$ (162 sq miles), serving 16 suburban cities in the Kansas City metropolitan area. Johnson County Wastewater operates a separate sanitary sewer system that provides wastewater collection and a total treatment capacity of approximately 2.85 m$^3$/s (65 mgd) at six regional WRRFs and 30 pumping stations. Johnson County Water's underground assets include 55,000 manholes and more than 3520 km (2200 miles) of pipe with an estimated replacement value of more than $1.5 billion.

Opportunities for cost-effective I/I reduction have focused on three main service areas located in the northeast portion of Johnson County. These areas represent the oldest portions of the JCW system and are largely made up of single-family residential homes with associated commercial and light industry development. Development of these areas occurred predominantly from 1940 to 1980. The public system in these areas is primarily comprised of vitrified clay pipe and brick manholes. These areas represent approximately 35% of JCW's total service area (1280 km [800 miles], 19,300 manholes).

## 2.2     Private Infiltration and Inflow Removal Program

The following subsections describe JCW's private I/I program. The case study presented in this publication is based on JCW's Pilot I/I Rehabilitation Project, which was conducted from 2009 to 2014. Information on JCW's historical program can be found in WEF's PPVL.

### 2.2.1   Program Drivers

Johnson County Wastewater's private I/I program is focused on two primary objectives: meeting regulatory requirements for sanitary sewer overflows (SSOs) and meeting customer levels of service for protection against basement backups during wet weather. Johnson County Wastewater's private I/I program began in the 1980s to address customer outcries caused by significant numbers of basement backups during large wet weather events. In the early 1980s, JCW's system recorded more than 300 wet weather backups per year (30 per 160 km [100 miles] of sewer). The need for improved system performance to protect customers from backups prompted action by JCW and county leaders to initiate a series of projects to address wet weather issues. Thus, JCW embarked on a $56 million improvement program in the late 1980s and 1990s. The private property I/I program was a key component of the overall improvement plan.

By the mid-2000s, with all of the planned system improvements completed, JCW system performance had improved significantly. Recorded backups for both dry weather and wet weather dropped notably and have consistently been below 20 per year (less than 1 per 160 km [100 miles] of sewer) in the last 5 years (i.e., 2011 to 2015).

Although I/I reduction efforts in the initial series of projects completed in the 1980s and 1990s were successful and system performance had dramatically improved, JCW still experienced high peak flows and system surcharging during significant wet weather events. In addition, JCW's oldest service area, served by its Nelson Wastewater Treatment Complex (hereafter referred to as the "Nelson WWTP Complex"), was designed and constructed with four satellite peak excess flow treatment facilities (PEFTFs) in the collection system for managing peak flows during wet weather. These facilities remain in service and discharge during large/intense wet weather events. In 2009, JCW responded to a U.S. Environmental Protection Agency (U.S.

EPA) 308(a) request for information regarding the current performance of these facilities and proposed alternatives for improving system performance related to wet weather discharges.

As part of the alternatives analysis in 2008 and 2009, JCW completed an evaluation that included cost-effectiveness analyses for addressing sewer overflows and basement backups that included estimates of the optimum level of I/I removal for the future. However, the initial analyses were based on information that had significant data gaps and assumptions, including the types of I/I sources that still exist in JCW's service area, the cost to remove these I/I sources, and the effectiveness of various I/I removal strategies in the JCW service area (including private lateral renewal).

As a result, JCW conducted the Nelson WWTP Complex Pilot I/I Rehabilitation Project from 2009 to 2014 to address these data gaps and assumptions by identifying the sources of I/I and evaluating the effectiveness and costs of several I/I removal strategies to reduce the risk of sewer overflows and basement backups within the two large sewersheds' tributary to the Nelson WWTP Complex.

### 2.2.2  Program Characteristics

The pilot project was designed to assess rehabilitation approaches and processes for I/I removal throughout the entire JCW service area by answering the following questions:

- How much I/I can JCW realistically expect to remove?
- What rehabilitation strategy is going to be effective, including upper and/or lower lateral rehabilitation? and
- What is the real cost of I/I removal?

One of the key objectives of this project was to evaluate and develop the processes and procedures for future removal of public and private sector I/I sources in the Nelson WWTP Complex service area and throughout the JCW service area.

The Nelson WWTP Complex service area is approximately 7284 ha (18 000 ac), with the pilot project area representing approximately 202 ha (500 ac) and 1,100 properties in two subsystems of the service area. The pilot area subsystems were selected because they are representative of the entire watershed, they had reasonably high levels of I/I, and because they are located in sections of the collection system that facilitated analysis of the effectiveness of I/I removal approaches.

To determine the most cost-effective I/I removal strategy, the two pilot project areas were divided into several smaller rehabilitation strategy areas in which different combinations of public and private sector I/I source removal were designed and constructed, including rehabilitation of JCW (public) sewer mains and manholes, removal of building and inflow sources, and rehabilitation of upper and/or lower laterals.

The results of the pilot project work will be used by JCW to implement a long-term and sustainable I/I reduction program using the lessons learned during the pilot project and identification of the recommended strategy. Implementation of the recommended I/I reduction strategy is scheduled to begin in 2016.

### 2.2.3 Public Outreach

In 2009, as JCW began the pilot I/I project as part of efforts to respond to a U.S. EPA 308(a) request for the Nelson WWTP Complex service area, the approach for public outreach related to building inspections and private I/I source removals was adjusted to accommodate the nature of the JCW pilot I/I project. In JCW's service area, the upper and lower laterals and connection of the laterals to the public main sewer are considered private property and were previously not addressed in the earlier private I/I program because of unknowns and limits in technology. Although JCW has the authority under its I/I code to require building inspections and the removal of illicit I/I connections on private property, a "voluntary approach" for the pilot I/I project was deemed more expedient than using a mandatory approach to compel compliance. As a result, the county's legal counsel developed property owner agreements as a means for JCW to remove illicit I/I connections at no direct expense to the property owner.

The success of the pilot I/I project was reliant upon the participation and acceptance of approximately 1,100 private property owners whose properties were affected by inspection and construction activities. It was critical for JCW and their project team to be effective in communicating with property owners to obtain the appropriate signed agreements so the investigation and construction activities could proceed. Therefore, a multifaceted public communication program was developed that included mailings, a project Web page, an I/I hotline phone and e-mail account, public meetings, individual property owner meetings, postcards and door hangers, follow-up telephone calls and door knocking efforts, and a survey.

During the investigation phase, building and lateral inspection participation exceeded 90% of targeted properties. During the source removal phase, building defect source removal and lateral rehabilitation participation exceeded 80% (see Table 5.1 for a summary of source removals).

The multifaceted public communication program used for the JCW pilot I/I project achieved the desired high property owner participation and satisfaction rates. The benefits of JCW's extensive public outreach efforts

**TABLE 5.1**    Private I/I investigation participation rates.

| Private building I/I removal | | | Upper and lower lateral service lines | | |
|---|---|---|---|---|---|
| Planned building I/I source removals | Signed building disconnect agreements | Percent of planned | Eligible properties | Signed service line rehabilitation agreements | Percent of eligible |
| 149 | 125 | 84% | 411 | 361 | 88% |

were twofold: reduced property owner complaints during construction and increased public support of the pilot I/I project. However, the level of effort required to implement a public outreach program of this nature must be considered against the desired participation and satisfaction rates. Indeed, significant resources were required to implement and execute the public outreach approach used by JCW for the pilot I/I project.

### 2.2.4   *Program Resources and Tools for Source Identification and Removal*

The following methods and resources were used to identify and remove private I/I sources:

- Inspections, consisting of smoke testing, dyed water testing, external and internal building inspections, service lateral closed-circuit television (CCTV) inspections, and nighttime flow monitoring;
- Source removals, consisting of area drains, stairwell drains, driveway drains, sump pumps connected to the sewer, sump pits and internal foundation drains, service lateral and connection repair/replacement using cured-in-place pipe, pipe bursting, and open-cut construction;
- Resources, consisting of
  - Consultants—engineering consultants were used predominantly to complete initial sewer system evaluation survey inspection work and determination of cost-effective I/I sources. For the 2009 through 2014 pilot I/I project, outside consultants were also used to complete the private I/I removal work by coordinating with the contractors to complete private service lateral rehabilitation and replacement and coordinating with the property owner and plumber for the private building I/I source removals;
  - Johnson County Wastewater staff—the JCW staff's role was primarily to administer the pilot I/I project and to support the consulting team members completing work in the field. Two JCW staff

members had significant involvement in the project to coordinate with the consulting team and plumbers on the private I/I activities associated with the building source removals;

○ Plumbers—private plumbing companies completed all of the I/I source removals. Plumbers were selected by property owners to complete I/I source removals on private property.

○ Funding—Johnson County approved funds for removal of private I/I sources at a set cost depending on the type of I/I source removal. Johnson County Wastewater coordinated with property owners and plumbers to either pay plumbers directly or to reimburse property owners for confirmed I/I source removals at the approved cost;

○ Johnson County risk management and legal staff—county risk management and legal staff provided legal support to JCW on all legal issues associated with the initial program. In the 1980s and 1990s, legal support was focused primarily on assisting with private property compliance with the I/I code and legal challenges or claims by private property owners against JCW. For the 2009 through 2015 pilot I/I project, legal support was focused primarily on assisting with development of agreements with private property owners to complete the work; and

○ Data management—for the 2009 through 2015 pilot I/I project, JCW's engineering consultants used databases and spreadsheets to track all the private I/I activities.

### 2.2.5  Program Costs

The pilot I/I project focused on two small areas that were part of the original I/I program with the intent of determining a cost-effective I/I strategy to define JCW's future program. The total construction cost of the pilot I/I project was $4.1 million, which included $390,000 for removal of 125 private I/I sources (average of $3,109 each). Approximately $2.34 million was spent to rehabilitate or replace 361 service laterals (average of $6,493 each) to assess the cost effectiveness of private I/I removal from upper and lower laterals. Note that some of the laterals had both the upper and lower lateral rehabilitated, while others had only the lower lateral rehabilitated.

### 2.3    Program Effectiveness

The pilot I/I project focused on two smaller areas ("MTM1" and "SMTC") within the Nelson WWTP Complex service area that were part of JCW's original I/I program. These two pilot areas were subdivided into various strategy areas to assess the effectiveness of different combinations of rehabilitation and repairs to accomplish I/I reduction. The following list describes the various strategy areas and work completed within these areas as part of the pilot I/I project:

- Everything—rehabilitation of all main sewers, manholes, service line connections, and upper and lower laterals, and removal of all identified building I/I sources;
- Private only—rehabilitation of all service line connections and upper and lower laterals, and removal of all identified building I/I sources;
- Public only—rehabilitation of all main sewers and manholes;
- Typical comprehensive—rehabilitation of main sewers and manholes contributing significant amounts of I/I, and removal of all identified building I/I sources/defects;
- Typical comprehensive plus lower lateral—rehabilitation of main sewers and manholes contributing significant amounts of I/I, removal of all identified building I/I sources, and rehabilitation of all service line connections and lower laterals; and
- Control—no rehabilitation work.

The I/I removal achieved within the MTM1 and SMTC pilot I/I areas was evaluated by comparing pre- and post-rehabilitation peak flows for the design condition. The percent I/I removal for each strategy is shown in Figure 5.1. The data show a range of removals from about 20 to 55% for the various rehabilitation strategies used. The "everything" strategy, where all mainline, manhole, service laterals, and buildings were rehabilitated regardless of condition, achieved a 55% I/I removal in MTM1 and a 47% I/I removal in SMTC. Even though all assets were rehabilitated, approximately 50% of the I/I was not removed, which may be attributable to active foundation drains, leaking basement walls, and defects in under-slab piping. In the "private" strategy area, all building sources were removed and all service laterals were rehabilitated regardless of condition. For this strategy area, 25% I/I reduction was achieved. In the "typical comprehensive and lower lateral" ("Typ + LL") strategy area, about 30% I/I was removed.

The capital cost to implement the rehabilitation in each strategy area was tracked so that the rehabilitation cost by rehabilitation strategy area and the I/I removal cost efficiency could be evaluated. These are shown in Figure 5.2.

The rehabilitation capital cost on a unit rate basis (i.e., capital cost per foot of main sewer) and the corresponding I/I reduction are shown in Figure 5.3. The data show that the "typical comprehensive" strategy used was the most effective in terms of cost and still achieves substantial I/I removal.

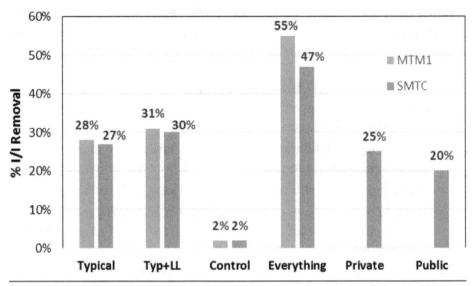

**FIGURE 5.1**  Johnson County Wastewater pilot I/I reduction by strategy area (JCW, 2014).

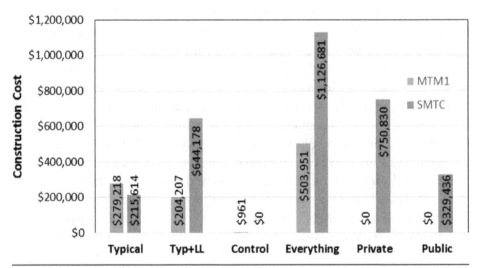

**FIGURE 5.2**  Johnson County Wastewater construction cost by strategy area (JCW, 2014).

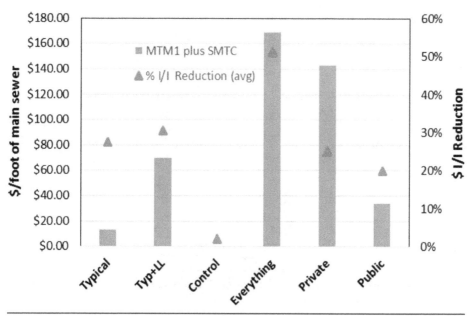

**FIGURE 5.3**    Johnson County Wastewater unit construction cost vs I/I reduction by strategy area (JCW, 2014).

This is further shown in Figure 5.4, which plots the dollars per gallon per day I/I removed in each strategy area. The "everything" strategy area is 4 times (i.e., the MTM1 pilot area) and 9 times (i.e., the SMTC pilot area) the cost per unit of I/I removed than the "typical comprehensive" strategy area.

## 2.4    Conclusions

The following are conclusions of JCW's pilot I/I project and considerations for the future program:

- Although JCW achieved significant I/I reductions in the past, I/I can still be cost-effectively reduced;
- To address I/I reduction for the long term, JCW had to commit time, energy, and resources to plan and execute the pilot I/I project to discover and develop the best long-term strategy for I/I reduction for the future. Past assumptions needed to be confirmed with real data and experience in JCW's system to confirm the best I/I reduction strategy for the future;

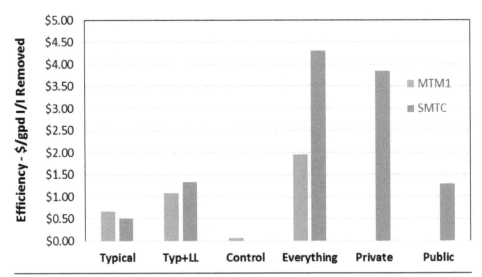

**FIGURE 5.4** Johnson County Wastewater I/I removal efficiency by strategy area (JCW, 2014).

- Reasonable predictions of I/I removal and costs can now be made to manage I/I reduction with the data collected and "real world" experience in JCW's system;

- The most effective I/I reduction strategy will include work in both the private and public sectors to achieve long-term cost-effective I/I goals. Johnson County Wastewater's strategy will include I/I source removal of sources that have traditionally been cost effective in the past in both the public sector (defective manholes and main lines) and private sector (i.e., sump pumps, interior foundation drains, etc.);

- Lower lateral rehabilitation should focus on cost-effective lateral repairs only. Based on the pilot I/I study, it was confirmed that repairing all lower laterals regardless of condition was not cost effective. However, data indicated that focusing on defective lower laterals may be cost effective. This needs to be confirmed with future work;

- Although the pilot I/I study provided good direction on the strategy for JCW's future I/I reduction efforts in both the public and private sector, the strategy will need to be implemented with a "continuous improvement" mindset. Flow monitoring before and after I/I reduction should be conducted to confirm effectiveness of the strategy and "lessons learned" should be evaluated to determine adjustments to the strategy in future basins to continue to maintain effectiveness of the program;

- Johnson County Wastewater has plans to implement and confirm the I/I reduction strategy recommended from the pilot I/I project. Implementation initially is planned for selected priority basins that have high peak wet weather flows and are upstream of areas with capacity issues; and

- Johnson County Wastewater intends to implement the future I/I reduction strategy with a long-term "program approach" to allow JCW to systematically address I/I reduction methodically over the long term.

# 3.0   KING COUNTY, WASHINGTON

King County, Washington, has been evaluating I/I removal since 1999, with recent I/I removal demonstration projects focused on evaluating the effectiveness of rehabilitation of upper and lower laterals.

## 3.1   Utility Background

King County, Washington, provides wholesale wastewater treatment services to 1.6 million people in 34 local sewer agencies in King, Snohomish, and Pierce counties. King County began serving the area in 1956. King County's wastewater system includes three large regional water resource recovery facilities (WRRFs), two small WRRFs, four combined sewer overflow (CSO) treatment facilities, 695.6 km (391 miles) of sewer lines, 25 regulator stations, 47 pumping stations, and 38 CSO outfalls.

King County Wastewater Treatment Division created the Regional I/I Control Program in 1999 as part of the Regional Wastewater Services Plan to explore the feasibility of I/I control. The county implemented multiple pilot projects in 2003 and 2004 to test various I/I reduction methods and technologies. Consensus recommendations from King County and local agencies were to implement and evaluate initial I/I reduction projects to test the cost effectiveness of I/I reduction on a larger scale than the pilot projects. Cost effectiveness was determined through a benefit-cost analysis that compared I/I reduction to other wastewater conveyance capacity improvements.

King County and the local agencies selected the Skyway Infiltration and Inflow Reduction Demonstration Project ("Skyway Project") to test the feasibility of large-scale I/I reduction. The Skyway Project is highlighted in this case study (King County Department of Natural Resources and Parks, 2014). Before the Skyway Project, 2003/2004 pilot projects were performed in a portion of the Skyway basin, resulting in an 88.5% reduction in peak I/I in the pilot basin.

## 3.2    Private Infiltration and Inflow Removal Program

The following subsections describe King County, Washington's, private I/I program.

### 3.2.1  Program Drivers

In the 1990s, flow estimates based on flow monitoring and population projections indicated the system would be out of capacity by the year 2010. The goal of the program is to reduce I/I to the sewer system, thereby increasing the capacity of the existing system and eliminating or reducing the need for capacity improvements. A key objective of the Skyway Project was to evaluate the effectiveness of sewer rehabilitation, specifically upper and lower laterals. The project goal was to determine if rehabilitation reduced I/I enough to delay, reduce the size of, or eliminate the need for a $11.3 million (2015 dollars) storage project.

### 3.2.2  Program Characteristics

The Skyway Project aimed to rehabilitate upper and lower laterals serving 343 of the 375 properties in the basin using pipe bursting with 10.2-cm- (4-in.-) diameter high-density polyethylene pipe. The project also included replacement of approximately 6000 m (20 000 ft) of primarily 20.3-cm- (8-in.-) diameter sewer mains by pipe bursting and replacement of 90 manholes. The Skyway Project was similar in many ways to the 2003/2004 pilot projects, including age of sewer pipe, materials used for rehabilitation, neighborhood characteristics, and condition of the sewers.

### 3.2.3  Public Outreach

One of the risks identified early on in the predesign project was that rights-of-entry would be attained for too few low- and medium-difficulty properties, requiring more work on high-difficulty projects and at a higher cost. If too few rights-of-entry were attained for the targeted amount of private property rehabilitation, the project could not proceed to implementation. Mailings were used to communicate the program to residents and were found to be a more effective method than the door-to-door approach.

Of the 375 properties identified for rehabilitation, 25 were located along a heavily traveled roadway and were omitted from the work because of the increased time and cost associated with working in heavy traffic areas. Ultimately, 343 properties were included for rehabilitation in the final design.

### 3.2.4  Source Identification and Removal

Upper and lower laterals were CCTV-inspected from the sewer main connection to the private property immediately before work on each property began. This allowed the sewer to be located on each property to check what was shown on the design. Once the line was located, pipe bursting pit locations were assessed and surface features that would be disturbed were evaluated. The CCTV also allowed for confirmation of service lateral materials and determined whether the line had already been replaced recently. Under the direction of King County inspectors, decisions on the extent of rehabilitation on each property were made following the CCTV work. Rehabilitation was performed within a day or two of the CCTV work. Rehabilitation was performed on 298 of the 343 properties included in the final design. The other 45 properties were omitted because of the upper and lower lateral being recently replaced with polyvinyl chloride (13 properties), surface features making rehabilitation risky (20 properties), property owners deciding against work because of landscape or hardscape being affected (four properties), or properties sharing a common service line, with one or more properties not being replaced for reasons previously stated (eight properties).

Table 5.2 shows the quantity of upper and lower laterals, sewer mains, and manholes that were bid and ultimately replaced. Less than 70% of the upper and lower lateral length that was bid was actually replaced. As shown in Table 5.2, 95% of the bid sewer main and 94% of the manhole replacements were constructed.

### 3.2.5  Program Resources and Tools

Various resources used during the project, including the Draft Regional I/I Control Standards and Policies, the right-of-entry form, intergovernmental

**TABLE 5.2**  King County Skyway Project rehabilitation quantities and costs (ft × 0.3048 = m).

| Description | Bid quantity | Average bid cost | Low bid cost | Final quantity |
|---|---|---|---|---|
| Entire laterals | 32 965 ft | $2,609,377 | $1,253,388 | 21 981 ft* |
| Sewer mains and manholes | 21 400 ft 99 manholes | $2,084,064 | $2,208,800 | 20 369 ft 93 manholes |
| Sales tax | | $445,877 | $311,808 | |
| Total | | $5,139,318 | $3,593,995 | |

*An additional 1300 ft of lateral was replaced by open cut.

agreements, and public education materials, can be found at http://king-county.gov/services/environment/wastewater/ii/resources.aspx; WEF's PPVL can be accessed at www.wef.org.

### 3.2.6 Program Costs

As shown in Table 5.2, the low bid price for the Skyway Project was $3,593,995. The low bid differed from the other bids in that it included more costs for the sewer main and manhole replacement and less costs for the upper and lower lateral replacement. The final construction cost was $3,417,626. The final cost included a change order for reduction in unit quantities ($311,830). The reduced quantities were primarily attributable to fewer upper and lower laterals being rehabilitated.

### 3.3    Program Effectiveness

Infiltration and inflow reduction was quantified by comparing model results based on flow data collection both pre- and post-rehabilitation. Data collected during the 2012/2013 wet-weather post-rehabilitation flow monitoring was input to the county's MOUSE hydrologic model to estimate the remaining I/I and to assess the effectiveness of I/I removal efforts.

The 2003/2004 pilot project conducted in a portion of the Skyway basin resulted in a peak flow reduction from 339 m³/h (2.15 mgd) to 39.4 m³/h (0.25 mgd) (89%). The county and local sewer agencies established a more conservative reduction target of 60% for the Skyway Project because achieving this would provide sufficient downstream system capacity to eliminate the need for the $11.3 million (2015 dollars) storage facility.

The I/I reduction effectiveness of the Skyway Project was less than the 2003/2004 pilot project in that the measured 19% I/I reduction was far below the expected 60% removal. As stated previously, the Skyway Project had strong similarities to the pilot project in that the neighborhood was the same age and similar construction materials, the same design, and the same pipe bursting replacement were used; additionally, the same contractor performed both projects and the same inspector observed both projects.

Factors that may have influenced the I/I reduction effectiveness include the following:

- Foundation drains and sump pumps might contribute flow to the sewer system that reduces the effectiveness of the rehabilitation efforts;
- Groundwater levels may become temporarily amplified in downslope areas after service lines no longer behave as groundwater drains. This could cause high infiltration pressures on unimproved facilities in the downstream portion of the project area; and

- Seventy-two properties replaced less than 75% of the service lines because of constructability constraints such as decks and patios.

In addition to these factors, it should be noted that a flow diversion occurred in the wastewater collection system during the project that likely confounded the analysis of reduction results.

## 3.4    Conclusions

Extensive work went into the development of the project, from the initial pilot project in 2003 through the postconstruction evaluation of the Skyway Project. The following lessons learned will help guide future I/I reduction projects in King County:

- Identifying problem areas—for residential areas, a general rule for a peak I/I flow equivalent of 11.4 L/m (3 gal/m) or more for each property is a good indicator of where to focus rehabilitation efforts;
- Basin characterization—lost or incomplete sewer inventory records can affect the ability to clearly define the sewer system configuration in a study area. A thorough assessment of physical conditions is important in defining basin boundaries and selecting flow meter locations for optimal hydraulics;
- Consideration of sump pumps and foundation drains—the likely effect of sump pumps on I/I removal effectiveness should be considered when high groundwater is documented in the area, the area has a high percentage of full or daylight basements, and with observations of clean, cold water discharge noted from the lateral when no one is home. Explore the possibility of routing foundation drains to the storm sewer system;
- Construction flexibility—implement construction with maximum potential for modifying design based on field conditions. Structure the contract based on unit prices to allow for the addition or deletion of work on properties;
- Groundwater issues—I/I reduction projects should anticipate the possibility of groundwater issues that may follow disturbed ground associated with pipe bursting and open trenching methods. Projects should consider groundwater interception and discharge methods for groundwater pressure relief;
- Contingency bid items—include contingency bid items for unexpected side sewer route adjustments, drainage problems, and landscape/hardscape restoration; and

- Upper and lower lateral replacement—to the extent it is cost effective, replace as much of the upper and lower service lateral as possible.

## 4.0   EAST BAY MUNICIPAL UTILITY DISTRICT

### 4.1   Utility Background

The East Bay Municipal Utility District (EBMUD) provides water and wastewater services to customers in the East Bay region of the San Francisco Bay. The EBMUD conveys and treats an annual average daily flow of approximately 2.9 m³/s (66 mgd) of wastewater generated by separate sewer systems in seven "satellite" agencies (six cities and one special district). The satellites include the cities of Alameda, Albany, Berkeley, Emeryville, Oakland, and Piedmont, and the Stege Sanitary District (serving a portion of the City of Richmond and the communities of Kensington and El Cerrito). The EBMUD's Interceptor System includes approximately 46.7 km (29 miles) of gravity sewers, 12.9 km (8 miles) of force mains, and 15 pumping stations; the satellite collection systems include approximately 2575 km (1600 miles) of sewers (gravity and force mains) and numerous pumping stations.

### 4.2   Private Infiltration and Inflow Removal Program

The following subsections present a summary of how EBMUD and its satellite communities implemented an ordinance for private sewer lateral replacement to achieve I/I reduction (Oriol et al., 2015).

#### 4.2.1   Program Drivers

The EBMUD Main Wastewater Treatment Plant provides primary treatment up to a capacity of 14.0 m³/s (320 mgd) and secondary treatment up to a capacity of 7.4 m³/s (168 mgd). The EBMUD also operates three PEFTFs, also referred to as "wet weather facilities", with a combined capacity of more than 13.1 m³/s (300 mgd) to store and/or treat and discharge peak flows. These facilities were constructed in the 1990s and operated under a National Pollutant Discharge Elimination System (NPDES) permit issued by the San Francisco Bay Regional Water Quality Control Board (RWQCB), which is the local NPDES-permitting authority pursuant to delegations by U.S. EPA and the California State Water Resources Control Board (SWRCB). However, in 2007, the SWRCB issued a remand of the NPDES permit for the EBMUD PEFTFs and directed the RWQCB to issue a new permit that either requires full secondary treatment of peak excess flows or prohibits discharge.

The East Bay system is unique in that there is no contractual or financial relationship between EBMUD and the satellite agencies. The EBMUD solely owns, operates, and maintains the interceptor and water resource recovery facilities (WRRFs) and bills customers directly for those services. Each satellite owns, operates, and maintains its collection system and bills customers directly for conveyance service. There are no restrictions on the amount of flow that enters EBMUD's system from the 162 satellite connection points and no direct financial incentives to limit flows. Therefore, although EBMUD was responsible for compliance with the obligation to cease discharges from the PEFTFs, compliance was not achievable without the cooperation of the satellites.

### 4.2.2  *Program Characteristics*

Under two interim stipulated orders (i.e., a 2009 stipulated order with EBMUD and 2011 stipulated orders with the satellites), and then under the 2014 regional consent decree, EBMUD and the satellites agreed to eliminate discharges from the PEFTFs by 2036. The consent decree requires that the defendants collectively reduce I/I sufficiently to eliminate discharges from the three PEFTFs under a design storm. This reduction will be achieved by addressing all three types of I/I: infiltration from main lines and manholes via sewer rehabilitation, I/I from private sewer laterals (PSLs) via PSL replacements, and public and private inflow identification and removal. The consent decree does this in a way that aligns consent decree work commitments with asset ownership. Furthermore, the consent decree is structured under an adaptive management approach that evaluates progress in PEFTF discharge reductions and allows for course correction during the term of the decree. If the benchmarks established by the consent decree are not met in a check-in year, the parties will conduct an evaluation to determine why progress did not meet expectations and whether any revisions to the work are required.

The consent decree also includes elements designed to reduce SSOs, such as cleaning, inspection, and root control programs. Approximately half the pipe length and potentially 50% of the infiltration in the East Bay regional system actually belong to property owners. Therefore, the parties agreed that the only way to reduce I/I sufficiently to cease wet weather discharges was to regulate PSLs. In EBMUD's service area, property owners own their PSLs from the structure to (and including) the connection to the sewer main. Under EBMUD's 2009 stipulated order, EBMUD was required to adopt and implement a regional PSL ordinance. The ordinance became effective in 2011 and requires property owners to certify their PSLs using a pressure test witnessed by EBMUD upon hitting one of the following three triggers:

- Transferring title (i.e., property sale);
- Performing remodeling or construction on a property in excess of $100,000; and
- Changing water meter size.

Because the majority of PSLs in EBMUD's service area are made of clay pipe that has deteriorated over the years, most PSLs require replacement to pass the pressure test. Property owners have the option to repair or simply test PSLs that are in good condition. The EBMUD issues compliance certificates valid for 20 years to properties where the PSL has been completely replaced. Where a repair has been done or the PSL passed "as is", a 7-year compliance certificate is issued. Since the inception of the program, EBMUD has certified more than 16,000 laterals, and the current compliance rate for the point-of-sale trigger is 96%.

The ordinance requirements apply to all property types, including residential, commercial, and industrial. Special provisions cover two unique sets of properties. For condominiums and other common interest developments where multiple owners share common laterals, compliance at title transfer is not required; however, all PSLs must be compliant by 2021, which is 10 years after ordinance adoption. This longer timeline allows homeowner associations to plan for and implement PSL rehabilitation projects over time. Similarly, parcels with more than 305 m (1000 ft) of PSLs (e.g., large school or industrial campuses) are required to submit a Condition Assessment Plan for their PSLs within 5 years, and then, subsequently, to submit a Corrective Action Plan outlining how they plan to bring PSLs into compliance. This provision was added for two reasons. First, these large campuses are less likely to transfer title and, therefore, would otherwise take longer to hit the triggers that would require rehabilitation. Second, these large systems are complex and often appear more like small city sewer systems. Therefore, additional time and resources are needed to bring them into compliance.

The EBMUD implements the regional ordinance in cooperation with six of the seven satellite agencies. These satellites permit the PSL work and inspect the PSLs for materials, methods, and connections to the sewer, while EBMUD conducts the verification of the pressure test and provides overall program management and enforcement. The City of Berkeley implements its own comparable ordinance, which was in place before EBMUD's ordinance. Since the inception of the regional PSL ordinance by EBMUD, 16,275 compliance certificates have been issued (36 in Albany, 158 in Alameda, 142 in Emeryville, 13,804 in Oakland, 724 in Piedmont, and 1411 in Stege). For fiscal year 2015, 74% of the PSLs investigated required replacement, 16% required repair, and 10% passed the verification test as is. The EBMUD checks property sale data against the compliance database monthly. Non-compliant properties receive two courtesy notices followed by a notice of violation. The compliance rate following the enforcement notices is greater than 90%. The PSL program staffing includes two wastewater control representatives, two administrative clerks (one office and one field), one senior field inspector, and five field inspectors.

The requirement for EBMUD and Berkeley to continue implementing their PSL ordinance requirements was carried over from the EBMUD and satellite stipulated orders into the consent decree. Although it was considered, the parties decided not to require a PSL grant or incentive program as part of the consent decree. This decision was based on lessons learned during EBMUD's implementation of the Private Lateral Incentive Program (PLIP), a requirement of the 2009 EBMUD stipulated order. The goal of the PLIP was to increase PSL repair and replacement rates above the rate achieved through the PSL ordinance triggers alone. The PLIP attempted to achieve its goal by providing property owners financial incentives to voluntarily perform sewer lateral work. The effectiveness of the PLIP, however, was low. As a result, the PLIP was not included as an element of the 2014 regional consent decree.

The EBMUD identified multiple reasons for the PLIP's ineffectiveness. First, the PLIP is largely redundant. Historical evidence indicates a majority of properties, including those offered incentives through the PLIP, will meet a PSL ordinance trigger before the end of the consent decree term. Consequently, the PLIP does not provide enhanced I/I reduction over that of the PSL ordinance and represents a redundant expense because any PSLs replaced through the PLIP would likely be replaced through the PSL ordinance triggers. A second reason the PLIP was ineffective is that participation rates remained stubbornly low. The EBMUD tried four different approaches to publicizing and distributing grant funds to eligible property owners, but public interest proved tepid.

Another limitation of the PLIP is its focus on sewer laterals. The EBMUD believes that although defective sewer laterals are an important source of the region's high I/I levels, inflow and rapid infiltration throughout the collection systems are also important. The PLIP's focus overlapped the focus of the PSL ordinance; both addressed sewer lateral defects. Achieving the consent decree's goals will likely require improvements in sewer lateral condition and elimination of inflow and rapid infiltration sources throughout the public and private portions of the regional wastewater collection system. Because a program to focus on inflow identification and removal complements the PSL ordinance in a way the PLIP did not, the 2014 consent decree did not include a PLIP requirement, but introduced a new inflow reduction program, the Regional Technical Support Program (RTSP). The RTSP, in combination with the PSL ordinance, will result in greater I/I reductions than the PLIP.

Field investigations will be ongoing throughout the RTSP to help locate specific sources of I/I. As required by the consent decree, each year EBMUD will give formal notification to each satellite of the identified I/I sources. The satellites will, in turn, incorporate I/I source information to their collection system capital improvement programs and develop customer communications programs for enforcement of private property inflow ordinance violations.

### 4.3    Program Effectiveness and Conclusions

The EBMUD and the satellite agencies will be required to prove, based on modeling efforts, that they have made sufficient progress toward the elimination of discharges from the PEFTFs. Failure to meet an interim benchmark will require EBMUD and the satellites to complete extensive flow monitoring and analysis to determine the causes of the missed targets. The adaptive management approach allows EBMUD and the satellites to work in a manner that is responsive to early results without overly prescriptive requirements for 22 years of system improvements. The check-in points provide the regulatory agencies the ability to track the progress of the program.

## 5.0   SUMMARY

The case studies presented in this chapter show a variety of approaches taken by utilities with different ownership and operation responsibilities. Many more examples demonstrating how different utilities are tackling I/I problems can be found in the WEF PPVL at www.wef.org. The Water Environment Research Foundation and U.S. EPA Web sites also document I/I removal case studies and highlight publications on the subject.

## 6.0   REFERENCES

Johnson County Wastewater (2014) *Collection System Analysis for Peak Excess Flow Treatment Facilities: Mission Township Main Sewer District No. 1 and Shawnee Mission Turkey Creek Watersheds—Supplement to January 2009 Technical Response to EPA 308(a) Request.*

King County Department of Natural Resources and Parks, Wastewater Treatment Division (2014) *Skyway Infiltration and Inflow Reduction Demonstration Evaluation Report;* Tetra Tech: Seattle, Washington.

Oriol, H. G.; Horenstein, B. K.; Zipkin, J. T.; Dinsmore, C. A.; Salmon, J. D. (2015) A Ground-Breaking New Wet Weather Consent Decree for East Bay Municipal Utility District and Its Satellite Communities. *Proceedings of the 88th Annual Water Environment Federation Technical Exhibition and Conference* [CD-ROM]; Chicago, Illinois, Sept 26–30; Water Environment Federation: Alexandria, Virginia.

# Appendix

# 2015 Water Environment Federation Private Property Infiltration and Inflow Survey

*Tyler Lewis*

| | | | |
|---|---|---|---|
| 1.0 SURVEY BACKGROUD | 91 | 3.0 SERVICE CONNECTIONS | 96 |
| 1.1 Survey Questions | 92 | 4.0 INFILTRATION AND INFLOW | 103 |
| 1.2 Survey Responses | 92 | 5.0 ENFORCEMENT AND FINANCING | 110 |
| 2.0 UTILITY CHARACTERISTICS | 92 | 6.0 MISCELLANEOUS | 115 |

## 1.0 SURVEY BACKGROUD

The Water Environment Federation (WEF) distributed an e-mail invitation to utility members in February 2015 to participate in a Web-based survey related to private property infiltration and inflow (I/I). The survey, which is termed the "2015 WEF Private Property I/I Survey", is structured to define utility characteristics followed by operational-specific questions to highlight critical data for industry knowledge sharing, learning opportunities, and key performance analyses.

This appendix provides the survey results and related data for all of the questions included in the survey. The appendix is designed to provide reference data for report analyses as well as external user use and analytics.

## 1.1    Survey Questions

The survey is comprised of 38 questions under the following five primary survey sections:

- Utility characteristics (eight questions),
- Service connections (12 questions),
- Infiltration and inflow (nine questions),
- Enforcement and financing (seven questions), and
- Miscellaneous (two questions).

## 1.2    Survey Responses

Response rates varied across the 38 questions; relative counts, ranges, and averages are provided for each question to inform the level of engagement and overall data quality for a given question. The total number of participants initiating the survey is 47 respondents, with average response counts in the mid-to-high thirties for most questions. For questions with open-ended responses, a minimum, maximum, and average value is provided.

The breadth of information requested in the survey aims to document the current state of private property I/I control such that comparative analyses can be conducted across time and geographical boundaries, yielding data for performance management measures and benchmarking.

Responses to each survey question are presented in the following sections.

## 2.0    UTILITY CHARACTERISTICS

| 1. Your utility is best described as: | | |
|---|---|---|
| Answer options | Response percent | Response count |
| Municipal/township/regional government | 88.6% | 39 |
| Special purpose district | 9.1% | 4 |
| Private/investor owned | 2.3% | 1 |
| Other (please specify) | | 3 |
| | *Answered question* | 44 |
| | *Skipped question* | 3 |

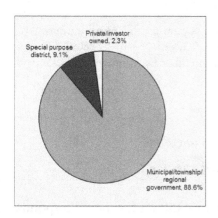

2. The length (in miles) of separated gravity sewers (excluding any satellite utility sewers owned and operated by another utility) in your collection system totals approximately how many miles (round to the nearest tenth of a mile)?

| | Response count |
|---|---|
| | 40 |
| *Answered question* | 40 |
| *Skipped question* | 7 |

The respondent with the longest collection system noted 2200 mi of separated gravity sewers. The respondent with the smallest reported 8.5 m of separated gravity sewers. The average length of gravity sewers for those respondents who provided the length of their gravity sewer system is 438 mi of separated gravity sewers.

3. The length (in miles) of combined gravity sewers in your collection system totals approximately how many miles (round to the nearest tenth of a mile)? Note: excluding any satellite utility sewers owned and operated by another utility.

| | Response count |
|---|---|
| | 37 |
| *Answered question* | 37 |
| *Skipped question* | 10 |

Of the six combined sewer system respondents, the largest respondent noted 219 mi of combined gravity sewers and the smallest respondent noted 5 mi of combined gravity sewers, with a survey average of 21 mi of combined gravity sewers.

4. The length of force mains in your collection system totals approximately how many miles (round to the nearest tenth of a mile)? Note: excluding any satellite utility force mains owned and operated by another utility.

|  | Response count |
|---|---|
|  | 40 |
| *Answered question* | 40 |
| *Skipped question* | 7 |

The length of force main varies from 0.1 mi to 450 mi for the 40 respondents that answered this question.

5. The length of low-pressure mains (serving nontraditional systems such as grinder pumps, vacuum sewer systems, etc.) in your collection system totals approximately how many miles (round to the nearest tenth of a mile)?

|  | Response count |
|---|---|
|  | 40 |
| *Answered question* | 40 |
| *Skipped question* | 7 |

Of the 21 respondents that provided a value for the length of low-pressure mains in their collection systems, the greatest amount is 38 mi of low-pressure mains and the smallest amount is 0.1 mi.

6. The service area served by your gravity sewer system totals approximately how many square miles (round to the nearest whole number)?

|  | Response count |
|---|---|
|  | 39 |
| *Answered question* | 39 |
| *Skipped question* | 8 |

The largest service area reported by a survey respondent is 600 sq mi. The smallest service area reported is 3 sq mi. Of the respondents that answered this question, the average size of the service areas is 81 sq mi.

**7. Does your collection system include any nontraditional collection system components (check all that apply)?**

| Answer options | Response percent | Response count |
|---|---|---|
| Grinder pump/pressure sewer systems | 67.6% | 23 |
| Vacuum sewer systems | 2.9% | 1 |
| Septic tank effluent pump (STEP) systems | 23.5% | 8 |
| No nontraditional components | 26.5% | 9 |
| Other (please specify) | | 2 |
| | *Answered question* | 34 |
| | *Skipped question* | 13 |

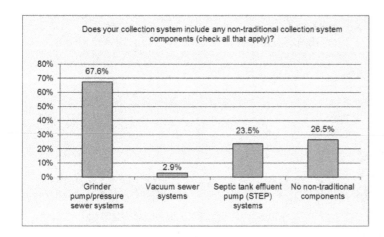

**8. The number of full time equivalent (FTE) staff dedicated to your collection system work, including management and support services, totals approximately (round to the nearest whole number):**

| | Response count |
|---|---|
| | 41 |
| *Answered question* | 41 |
| *Skipped question* | 6 |

The number of FTEs from respondents who answered this question ranged from two to 150, with the average number of FTEs equaling 31.

## 3.0   SERVICE CONNECTIONS

**9.** The estimated population served by your collection system totals approximately? Note: excluding populations served by satellite utilities with their own collection systems.

| Answer options | Response percent | Response count |
|---|---|---|
| <10,000 people | 17.9% | 7 |
| 10,000 to 100,000 people | 46.2% | 18 |
| 100,000 to 500,000 people | 28.2% | 11 |
| 500,000 to 1,000,000 people | 5.1% | 2 |
| >1,000,000 people | 2.6% | 1 |
| | *Answered question* | 39 |
| | *Skipped question* | 8 |

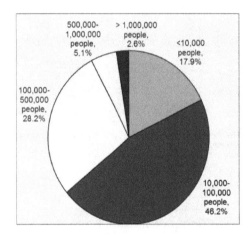

**10.** How many active service taps are currently installed in your collection system? Please specify the number:

| | Response count |
|---|---|
| | 36 |
| *Answered question* | 36 |
| *Skipped question* | 11 |

The largest respondent noted 137,031 taps and the smallest respondent noted 255, with a survey average of 28,426 taps.

**11.** With respect to active service taps on the sewer main, what percentage is located and mapped?

| Answer options | Response percent | Response count |
| --- | --- | --- |
| 90 to 100% | 48.6% | 18 |
| 50 to 89% | 18.9% | 7 |
| 11 to 49% | 16.2% | 6 |
| 0 to 10% | 16.2% | 6 |
| | *Answered question* | 37 |
| | *Skipped question* | 10 |

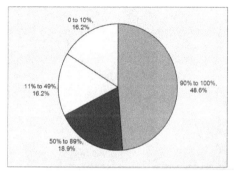

**12.** With respect to the lower laterals, what percentage is located and mapped?

| Answer options | Response percent | Response count |
| --- | --- | --- |
| 90 to 100% | 37.8% | 14 |
| 50 to 89% | 18.9% | 7 |
| 11 to 49% | 13.5% | 5 |
| 0 to 10% | 29.7% | 11 |
| | *Answered question* | 37 |
| | *Skipped question* | 10 |

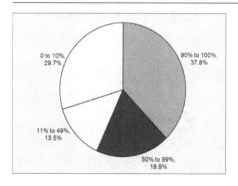

**13. With respect to the upper laterals, what percentage is located and mapped?**

| Answer options | Response percent | Response count |
| --- | --- | --- |
| 90 to 100% | 18.9% | 7 |
| 50 to 89% | 18.9% | 7 |
| 11 to 49% | 10.8% | 4 |
| 0 to 10% | 51.4% | 19 |
| | *Answered question* | 37 |
| | *Skipped question* | 10 |

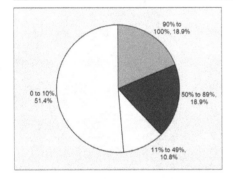

**14. How does your utility distinguish publicly owned and maintained laterals vs privately owned and maintained laterals?**

| Answer options | Response percent | Response count |
| --- | --- | --- |
| Public assets include the sewer main (typically in the street) only and exclude the service tap. | 31.6% | 12 |
| Public assets include the sewer main (typically in the street) and the service tap at the main. | 18.4% | 7 |
| Public assets include the sewer main (typically in the street), the service tap at the main, and the lower lateral pipe to the street right-of-way or easement line. | 47.4% | 18 |
| Public assets include the sewer main (typically in the street), the service tap at the main, the lower lateral, and the upper lateral from the street right-of-way or easement line to the building cleanout. | 2.6% | 1 |
| Other (please specify) | | 1 |
| | *Answered question* | 38 |
| | *Skipped question* | 9 |

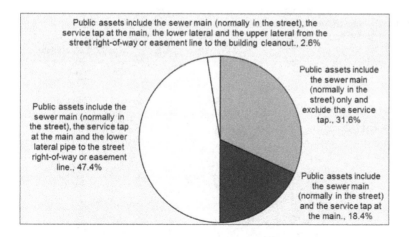

**15.** Please check all that apply relative to your utility's current cleanout policy and practice:

| Answer options | Required | Typically exists | No cleanout required | Response count |
|---|---|---|---|---|
| At building | 16 | 14 | 7 | 34 |
| At right-of-way | 13 | 8 | 16 | 32 |
| At easement | 7 | 4 | 20 | 29 |
| | | | *Answered question* | 37 |
| | | | *Skipped question* | 10 |

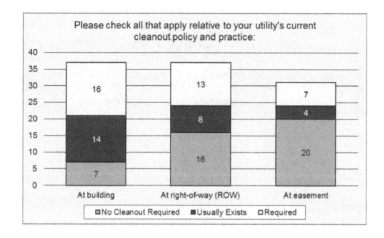

**16. Does your utility have an inspection program for service taps (check all that apply)?**

| Answer options | Response percent | Response count |
|---|---|---|
| Yes, for new installations | 81.1% | 30 |
| Yes, for repairs | 67.6% | 25 |
| Yes, when performing I/I or other investigations | 51.4% | 19 |
| Yes, upon property ownership transfer | 2.7% | 1 |
| No | 2.7% | 1 |
| Other (please specify) | | 4 |
| | *Answered question* | 37 |
| | *Skipped question* | 10 |

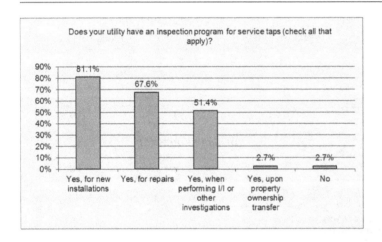

**17. Does your utility have an inspection program for lower laterals (check all that apply)?**

| Answer options | Response percent | Response count |
|---|---|---|
| Yes, for new installations | 67.6% | 25 |
| Yes, for repairs | 59.5% | 22 |
| Yes, when performing I/I or other investigations | 43.2% | 16 |
| Yes, upon property ownership transfer | 0.0% | 0 |
| No | 10.8% | 4 |
| Other (please specify) | | 4 |
| | *Answered question* | 37 |
| | *Skipped question* | 10 |

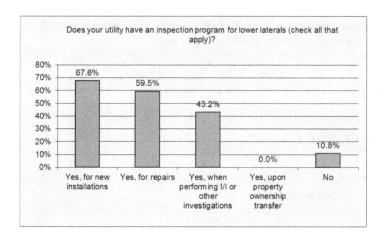

## 18. Does your utility have an inspection program for upper laterals (check all that apply)?

| Answer options | Response percent | Response count |
|---|---|---|
| Yes, for new installations | 56.8% | 21 |
| Yes, for repairs | 37.8% | 14 |
| Yes, when performing I/I or other investigations | 21.6% | 8 |
| Yes, upon property ownership transfer | 5.4% | 2 |
| No | 29.7% | 11 |
| Other (please specify) | | 6 |
| | *Answered question* | 37 |
| | *Skipped question* | 10 |

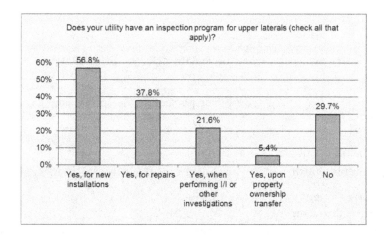

**19. Does your utility perform operation and maintenance services on the privately owned portion of the lateral (e.g., clearing blockages, root removal, etc.)?**

| Answer options | Response percent | Response count |
|---|---|---|
| Never, no legal authority to do so | 65.7% | 23 |
| No, have legal authority to do so, but is done to maintain good customer relationships because our crew is already at the site | 17.1% | 6 |
| No, but have legal authority if we decide to change policy and require repair/replacement | 17.1% | 6 |
| Yes, legal authority to do so | 0.0% | 0 |
| Other/comment (please specify) | | 5 |
| | *Answered question* | 35 |
| | *Skipped question* | 12 |

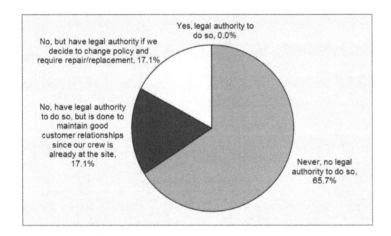

**20. Does your utility require property owners to repair or replace defective private lateral pipes and cleanouts?**

| Answer options | Response percent | Response count |
|---|---|---|
| Never, no legal authority to do so | 16.2% | 6 |
| No, but have legal authority if we decide to change policy and require repair/replacement | 10.8% | 4 |
| No, but we include private property lateral repairs as part of our sewer main improvement projects | 5.4% | 2 |

| | | |
|---|---|---|
| Yes, but only occasionally exercised | 43.2% | 16 |
| Yes, and routinely implemented when needed | 24.3% | 9 |
| Other/comment (please specify) | | 5 |
| | *Answered question* | 37 |
| | *Skipped question* | 10 |

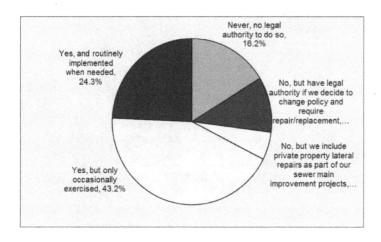

## 4.0   INFILTRATION AND INFLOW

**21. During dry, average, and wet years, what percent of total flow is I/I-related (check one response for each row)?**

| Answer options | <20% | 20 to 40% | 40 to 60% | 60 to 80% | >80% | Unsure | Response count |
|---|---|---|---|---|---|---|---|
| During a dry year | 29 | 3 | 0 | 0 | 0 | 3 | 35 |
| During an average year | 9 | 21 | 2 | 0 | 0 | 3 | 35 |
| During a wet year | 2 | 10 | 14 | 4 | 0 | 5 | 35 |
| | | | | | *Answered question* | | 35 |
| | | | | | *Skipped question* | | 12 |

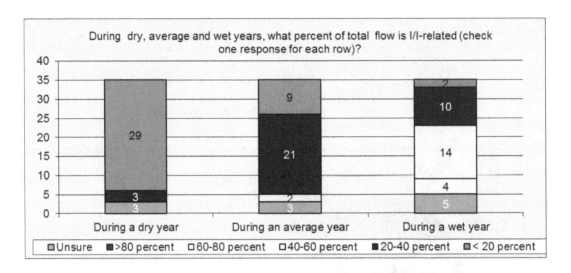

**22. What percent of total I/I do you estimate is related to building source connections, private laterals, and service taps?**

| Answer options | Response percent | Response count |
|---|---|---|
| <5% | 5.6% | 2 |
| 5 to 20% | 19.4% | 7 |
| 20 to 50% | 36.1% | 13 |
| 50 to 75% | 30.6% | 11 |
| >75% | 0.0% | 0 |
| Unknown | 8.3% | 3 |
| | *Answered question* | 36 |
| | *Skipped question* | 11 |

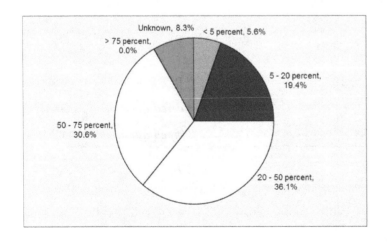

**23.** The percentage of total I/I related to building source connections, private laterals, and service taps indicated in the previous question is based on:

| Answer options | Response percent | Response count |
|---|---|---|
| A guess | 60.6% | 20 |
| Estimated source defect I/I unit contribution rates from a sewer system evaluation survey (SSES) | 27.3% | 9 |
| Basin flow monitoring | 12.1% | 4 |
| Lateral or service connection flow monitoring | 0.0% | 0 |
| Other/comment (please specify) | | 5 |
| | *Answered question* | 33 |
| | *Skipped question* | 14 |

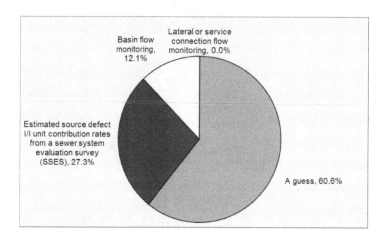

**24.** Does your utility monitor, measure, or quantify I/I from private I/I sources?

| Answer options | Response percent | Response count |
|---|---|---|
| Yes | 5.7% | 2 |
| No | 54.3% | 19 |
| Only as part of sewer system evaluation surveys (SSES) or other I/I studies | 40.0% | 14 |
| | *Answered question* | 35 |
| | *Skipped question* | 12 |

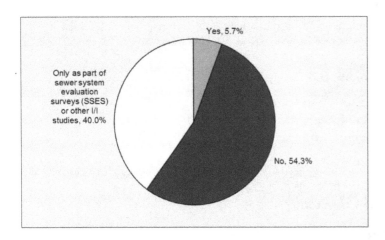

**25. If the answer to the previous question is "yes", would your utility be willing to share your methods, procedures, and results?**

| Answer options | Response percent | Response count |
|---|---|---|
| No | 55.6% | 5 |
| Yes | 44.4% | 4 |
| If yes, please provide contact name and information: | | 4 |
| | *Answered question* | 9 |
| | *Skipped question* | 38 |

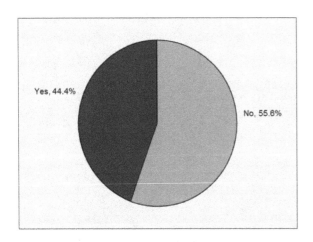

**26. What criteria are used for ranking for service area basins for I/I correction, rehabilitation, or remediation? (Assign ranking 1 as top priority and ranking 8 as last priority.)**

| Answer options | 1 | 2 | 3 | 4 | 5 | 6 | 7 | 8 | Rating average | Response count |
|---|---|---|---|---|---|---|---|---|---|---|
| Gross flowrates | 0 | 3 | 6 | 11 | 5 | 4 | 2 | 1 | 4.34 | 32 |
| I/I flowrates | 15 | 3 | 2 | 1 | 6 | 1 | 4 | 0 | 2.97 | 32 |
| Pump operating times and/or alarm conditions | 3 | 5 | 6 | 2 | 3 | 6 | 5 | 2 | 4.41 | 32 |
| Building backups | 2 | 8 | 3 | 6 | 1 | 4 | 4 | 4 | 4.38 | 32 |
| Sanitary or combined sewer overflows (CSOs) (sanitary sewer overflows or CSOs) | 8 | 7 | 6 | 4 | 3 | 0 | 2 | 2 | 3.16 | 32 |
| Growth/ availability capacity | 0 | 1 | 2 | 4 | 3 | 8 | 4 | 10 | 6.09 | 32 |
| Regulatory enforcement action or pressure | 2 | 3 | 3 | 3 | 7 | 3 | 9 | 2 | 5.03 | 32 |
| National Pollutant Discharge Elimination System permit requirements | 2 | 2 | 4 | 1 | 4 | 6 | 2 | 11 | 5.63 | 32 |

*Answered question*    32

*Skipped question*    15

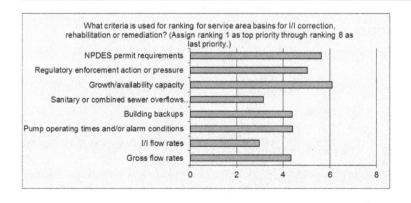

**27. Which I/I evaluation methods are used by your utility for private property (check all that apply)?**

| Answer options | Response percent | Response count |
|---|---|---|
| Nighttime flow monitoring | 28.6% | 10 |
| Wet weather flow monitoring | 48.6% | 17 |
| Pre- and post-rehabilitation flow monitoring | 40.0% | 14 |
| Closed-circuit television inspection of laterals | 60.0% | 21 |
| Smoke testing | 68.6% | 24 |
| Dyed water flood testing | 48.6% | 17 |
| Internal building inspection | 42.9% | 15 |
| None | 11.4% | 4 |
| | *Answered question* | 35 |
| | *Skipped question* | 12 |

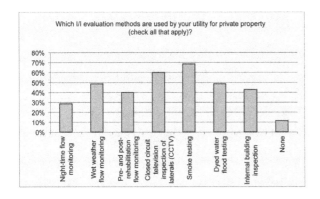

**28. What are the sources/causes of private I/I in your collection system? (Assign ranking 1 as the highest entry source and ranking 4 as the lowest entry source.)**

| Answer options | 1 | 2 | 3 | 4 | Rating average | Response count |
|---|---|---|---|---|---|---|
| Building sources | 14 | 7 | 9 | 4 | 2.09 | 34 |
| Upper lateral | 11 | 13 | 6 | 4 | 2.09 | 34 |
| Lower lateral | 7 | 12 | 13 | 2 | 2.29 | 34 |
| Cleanouts | 2 | 2 | 6 | 24 | 3.53 | 34 |
| | | | | | *Answered question* | 34 |
| | | | | | *Skipped question* | 13 |

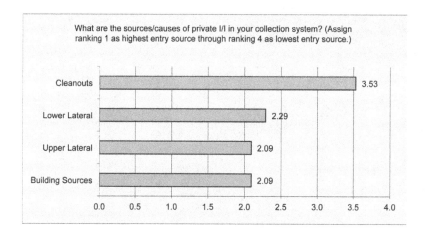

What are the sources/causes of private I/I in your collection system? (Assign ranking 1 as highest entry source through ranking 4 as lowest entry source.)

| | |
|---|---|
| Cleanouts | 3.53 |
| Lower Lateral | 2.29 |
| Upper Lateral | 2.09 |
| Building Sources | 2.09 |

**29. Which rehabilitation techniques and procedures are performed by your utility to address private I/I (check all that apply)?**

| Answer options | Response percent | Response count |
|---|---|---|
| None | 25.7% | 9 |
| Lateral point repair or replacement | 42.9% | 15 |
| Service tap robotic point repair | 11.4% | 4 |
| Sewer main lining with service tap liner | 42.9% | 15 |
| Lateral and service tap lining | 22.9% | 8 |
| Lateral joint test and grout | 8.6% | 3 |
| Lateral pipe bursting | 2.9% | 1 |
| Disconnecting area drain (driveway, stairwell, yard, window well) | 48.6% | 17 |
| Disconnecting roof drains | 45.7% | 16 |
| Disconnecting sump pumps/pits | 40.0% | 14 |
| Disconnecting foundation drains | 40.0% | 14 |
| Backflow preventer installation | 31.4% | 11 |
| Other (please specify) | | 7 |
| *Answered question* | | 35 |
| *Skipped question* | | 12 |

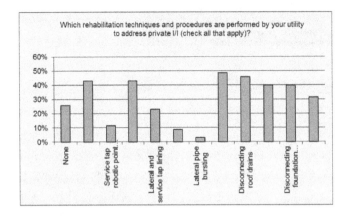

# 5.0   ENFORCEMENT AND FINANCING

30. **Does your entity require property owners to disconnect sump pumps/downspout/roof drain/area drain connections to the sanitary sewer system?**

| Answer options | Response percent | Response count |
| --- | --- | --- |
| Never, none of these prohibited connections exist in our collection system | 0.0% | 0 |
| Never, no legal authority to do so | 0.0% | 0 |
| No, but have legal authority if we decide to change policy to require such disconnection | 5.7% | 2 |
| Yes, but only occasionally exercised | 37.1% | 13 |
| Yes, and routinely implemented when needed | 57.1% | 20 |
| Unknown | 0.0% | 0 |
| | *Answered question* | 35 |
| | *Skipped question* | 12 |

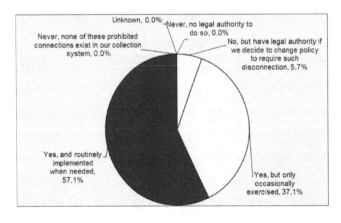

**31. If a prohibited connection is identified, how does your utility enforce the disconnection requirement (check all that apply)?**

| Answer options | Response percent | Response count |
|---|---|---|
| Written notification followed by a fine if not corrected | 84.6% | 22 |
| Penalty added to customer bill until connection removed | 15.4% | 4 |
| Lien against the property | 7.7% | 2 |
| Disconnect water service | 23.1% | 6 |
| Nothing | 15.4% | 4 |
| Other (please specify) | | 11 |
| | *Answered question* | 26 |
| | *Skipped question* | 21 |

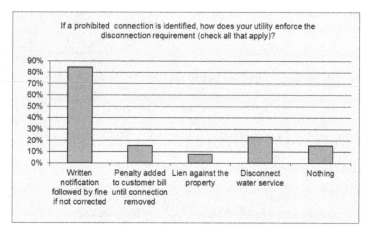

**32. Do you revisit the site and/or building at a later date to confirm the disconnection?**

| Answer options | Response percent | Response count |
|---|---|---|
| Yes | 78.8% | 26 |
| No | 21.2% | 7 |
| | *Answered question* | 33 |
| | *Skipped question* | 14 |

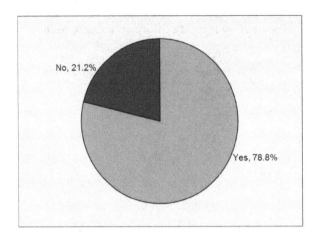

**33. If you answered "yes" to the previous question, how long do you wait before you revisit the site (indicate time in number of months)?**

|  | Response count |
| --- | --- |
|  | 26 |
| *Answered question* | 26 |
| *Skipped question* | 21 |

Of the 26 respondents inspecting sites to confirm disconnection, the average time is 2 months, with responses averaging from 0.5 months to 3 months.

**34. If your utility rehabilitates private laterals or service taps, how is the rehabilitation financed?**

| Answer options | Response percent | Response count |
| --- | --- | --- |
| Utility pays | 47.4% | 9 |
| Customer pays | 52.6% | 10 |
| Other or combination (please specify) |  | 11 |
|  | *Answered question* | 19 |
|  | *Skipped question* | 28 |

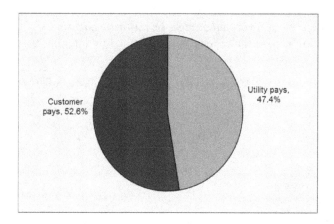

**35.** **If the customer pays for private lateral or service tap rehabilitation indicate the arrangements made for payment.**

| Answer options | Response percent | Response count |
|---|---|---|
| Customer pays full cost | 95.2% | 20 |
| Customer pays up to a maximum amount | 0.0% | 0 |
| Customer payment plans available | 0.0% | 0 |
| Customer payment plans available only to low or moderate income customers | 4.8% | 1 |
| Property liens | 0.0% | 0 |
| Insurance plan | 0.0% | 0 |
| Other (please specify) | | 7 |
| | *Answered question* | 21 |
| | *Skipped question* | 26 |

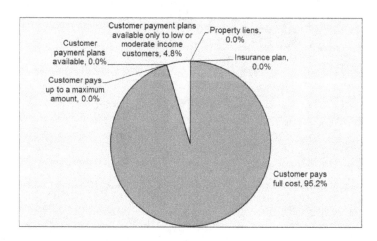

**36. Has your utility undertaken any measures to force, encourage, or motivate customers to remove prohibited connections or otherwise make repairs to their private laterals or service taps?**

| Answer options | Response percent | Response count |
| --- | --- | --- |
| No such measures undertaken | 38.7% | 12 |
| Mandatory insurance-type program to assist in paying for private repair work | 0.0% | 0 |
| Voluntary incentive program to assist in paying for private repair work | 12.9% | 4 |
| Enforcement-driven program to require property owners to pay for their private repair work | 35.5% | 11 |
| My utility pays for private repair work when we require such work to be performed | 12.9% | 4 |
| Other (please specify) | | 6 |
| | *Answered question* | **31** |
| | *Skipped question* | **16** |

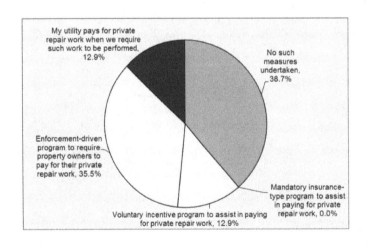

## 6.0   MISCELLANEOUS

**37. Do you have written procedures and technical specifications for the proper connection of a building to a public sewer line?**

| Answer options | Response percent | Response count |
|---|---|---|
| Yes | 94.3% | 33 |
| No | 5.7% | 2 |
| | *Answered question* | 35 |
| | *Skipped question* | 12 |

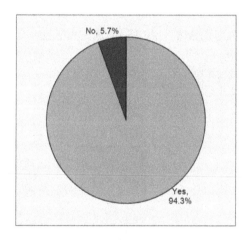

**38. Does your utility have a formal lateral cross bore inspection program in place?**

| Answer options | Response percent | Response count |
|---|---|---|
| Yes | 12.1% | 4 |
| No | 87.9% | 29 |
| Other (please specify) | | 2 |
| | *Answered question* | 33 |
| | *Skipped question* | 14 |

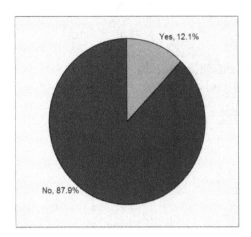

Questions and additional comments, information, or data, should be directed to:

Jane McLamarrah, Task Force Chair
Jane.Mclamarrah@mwhglobal.com
214.697.5567

# Index

**A**

Adaptive management, 52
Asset management software, 30
    building inspection modules, 30
    lateral modules, 30

**B**

Basin monitoring, 21
Best management practices (BMPs), 32
Budgeting, 51
Building inspections, 24

**C**

Capacity, management, operation and
        maintenance (CMOM)
        guidance, 15
Case studies, 69
    East Bay Municipal Utility District,
        84
    Johnson County Water, Kansas, 69
    King County, Washington, 79
Closed-circuit television, 25, 30
    cleanout entry, 25
    from main sewer, 26
Construction standards and
        specifications, 49
*Control of Infiltration and Inflow in*
        *Private Building Sewer*
        *Connections,* 2, 5
Corrective action, 30
    program, 60
    preventive design methods, 30
Corrective design methods
    cleanout repairs, 33
    downspout, driveway drain, and
        area drain removals, 33
    redirecting source flows, 35

    stairwell drain removals, 33
    sump pumps and foundation drain
        removals, 34
Customer, equity, 56
Customers, vulnerable, 54

**D**

Data, removal, 38
Default flow method, 27
Dyed water testing, 25, 28

**E**

East Bay Municipal Utility District, 84
Economic issues, 14
Electro-scan testing, 26
Enforcement efforts, 16, 110
    legal authority, 16
    legal requirements, 16
Enforcement-based program, 59
Environmental and public health
        issues, 15
Environmental justice, 56
Equity, customer, 56

**F**

Field verification of source flows
        method, 28
Financing, 110
Flow chart, point-of-sale, 62
Flow monitoring, 21
    Methods, 21
        basin monitoring, 21
        manual manhole service lateral
            monitoring, 21
        service lateral monitoring, 22
    postconstruction, 38
    preconstruction, 38

Flowrates
  establishing, 27
    default flow method, 27
    field verification of source
      flows method, 28
    surface runoff method, 27
Funding, 52
  allocation, 54
  effects, 55
  mechanisms, 56
  requirements, 55
Funds, public, 53

**G**
Geographical information system
  (GIS), 29

**I**
I/I initiatives
  education, 37
  public information, 37, 72
  effectiveness, 38
I/I source, 10, 20
  data management, 29, 74
    asset management software, 30
    geographical information
      system, 29
    private sector, 29
  inside, 28
  outside, 28
I/I source identification, 20, 23
  building inspections, 24
  dyed water testing, 25
  electro-scan testing, 26
  service lateral closed-circuit
    television, 25, 81
  smoke testing, 23
I/I magnitude, 12, 22
  estimating, 22
  flow monitoring, 21
I/I extent, 10
Information management, 51
Inspection programs, 9

**J**
Johnson County Water, Kansas, 69

**K**
Key performance indicators (KPIs), 52
King County, Washington, 79

**L**
Lateral installation
  terminology, 7
  typical, 7
Legal authority
  construction standards and
    specifications, 49
  Sewer Use Ordinance, 47

**M**
Manual manhole service lateral
  monitoring, 21

**N**
Nelson Wastewater Treatment Plant, 71

**O**
Operation and maintenance, 61

**P**
Performance metrics, 52
Point-of-sale
  flow chart, 62
  lateral inspection program, 60
Preventive design methods
  new sewer design, 30
  private connections to existing
    lines, 30
Private infiltration and inflow, effects,
  14
Private lateral repair methods, 35
  complete liner, 35
  grouting, 36
  partial liner, 35
  replacement, 36
    open cut, 36
    pipe bursting, 36
Private property program
  acceptable technologies, 52
  adaptive management, 52
  budgeting, 51
  costs, 74, 81

education, 50
effectiveness, 74, 81, 88
elements of, 46
example, 58
    City of San Antonio, 64
    Grand Strand Water and Sewer
        Authority, 63
    San Bruno, California, 60
funding, 52
information management, 51
legal authority, 47
management, 49
performance metrics, 52
politics, 56, 58
public education, 50
regulations, 56, 58
scope, 46
staffing, 49, 50
stakeholder involvement, 42
standard practices, 52
sustainability, 52
types of, 58
    corrective action program, 60
    enforcement-based, 59
    operation and maintenance,
        61, 79
    point-of-sale lateral inspection
        program, 60, 86
    publicly owned (lower) lateral
        focused program, 64
    utility-assumed ownership
        program, 61
vision, 46
Private Property Virtual Library
    (PPVL), 4, 46, 68
Private property rights, 57
Public outreach, 72
Public vs private
    obligations, 6
    responsibilities, 6
Publicly owned (lower) lateral focused
    program, 64

**R**
Regulatory issues, 15, 56

Removal data, 38
Removal programs, 37, 70
Repair methods, private lateral, 35
Rights, private property, 57

**S**
Service lateral closed-circuit television,
    25
Service lateral monitoring, 22
Sewer Use Ordinance, 47
Skyway, 79
Smoke testing, 23
Source flows, redirecting, 35
Source identification
    removal, 73
    tools, 73
Source testing
    postconstruction, 38
    preconstruction, 38
Staffing, 50
Stakeholder
    customer, 44
    internal, 44
    involvement, 42
    other, 45
    plumbers, service and repair
        contractors, builders,
        44
    real estate industry, 45
Sump pumps, 34
Sump pumps, data loggers, 29
Surface runoff method, 27
Survey, WEF 2015, 3, 91
Sustainability, 52

**T**
Terminology, 6
Terminology, lateral installation, 7

**U**
U.S. EPA, 43
Utility capital improvement funding,
    55
Utility-assumed ownership program,
    61

CPSIA information can be obtained
at www.ICGtesting.com
Printed in the USA
LVHW021738191220
674585LV00001B/12

9 781572 783270